Acoustic, Electromagnetic, Neutron Emissions from Fracture and Earthquakes

Alberto Carpinteri • Giuseppe Lacidogna
Amedeo Manuello

Editors

Acoustic, Electromagnetic, Neutron Emissions from Fracture and Earthquakes

 Springer

Editors
Alberto Carpinteri
Department of Structural, Geotechnical
and Building Engineering
Politecnico di Torino
Torino, Italy

Giuseppe Lacidogna
Department of Structural, Geotechnical
and Building Engineering
Politecnico di Torino
Torino, Italy

Amedeo Manuello
Department of Structural, Geotechnical
and Building Engineering
Politecnico di Torino
Torino, Italy

ISBN 978-3-319-16954-5   ISBN 978-3-319-16955-2   (eBook)
DOI 10.1007/978-3-319-16955-2

Library of Congress Control Number: 2015954667

Springer Cham Heidelberg New York Dordrecht London

Printed on acid-free paper

Springer International Publishing AG Switzerland is part of Springer Science+Business Media (www.springer.com)

Foreword

This book presents the relevant consequences of recently discovered and interdisciplinary phenomena triggered by local mechanical instabilities.

Vibrations at the TeraHertz frequency are in fact produced at the nano-scale in solids and fluids by fracture and cavitation, respectively. They present a frequency close to that of resonance in the atomic lattices and an energy close to that of thermal neutrons. A series of fracture experiments on natural rocks demonstrate that the TeraHertz vibrations are able to induce fission reactions on medium weight elements accompanied by neutron emissions, without gamma radiation and radioactive wastes.

The same phenomenon appears to have occurred in several different situations and, in particular, in the chemical evolution of Earth and Solar System, through seismicity (rocky planets) and storms (gaseous planets). It can also explain puzzles related to the history of our planet, like the ocean formation or the primordial carbon pollution, as well as scientific mysteries, like the so-called cold nuclear fusion or the correct radio-carbon dating of organic materials.

Very important applications to earthquake precursors, climate change, energy production, and cellular biology cannot be excluded in the future on the basis of the topics treated in the present volume. Scientists engaged in seismology, geophysics, geochemistry, climatology, planetology, condensed matter physics, biology, besides theoretical and applied mechanics, could receive a great benefit from the innovative and holistic view of this book.

I would like to thank my two Co-Editors, Prof. Giuseppe Lacidogna and Dr. Amedeo Manuello, for their untiring efforts during the last seven years in carrying out with me this difficult and cutting-edge research work.

A sincere and thankful acknowledgement is also due to the Co-Authors of the different chapters for their important scientific contributions. Thanks are in particular due to Dr. Oscar Borla for his specific expertise and contribution in environment neutron measurements.

Torino, Italy
September 2015

Alberto Carpinteri

Contents

1 TeraHertz Phonons and Piezonuclear Reactions from Nano-scale Mechanical Instabilities 1
Alberto Carpinteri

2 Correlation Between Acoustic and Other Forms of Energy Emissions from Fracture Phenomena 11
Giuseppe Lacidogna, Oscar Borla, Gianni Niccolini, and Alberto Carpinteri

3 Neutron Emissions and Compositional Changes at the Compression Failure of Iron-Rich Natural Rocks 23
Amedeo Manuello, Riccardo Sandrone, Salvatore Guastella, Oscar Borla, Giuseppe Lacidogna, and Alberto Carpinteri

4 Frequency-Dependent Neutron Emissions During Fatigue Tests on Iron-Rich Natural Rocks 39
Alberto Carpinteri, Francesca Maria Curà, Raffaella Sesana, Amedeo Manuello, Oscar Borla, and Giuseppe Lacidogna

5 Alpha Particle Emissions from Carrara Marble Specimens Crushed in Compression and X-ray Photoelectron Spectroscopy of Correlated Nuclear Transmutations 57
Alberto Carpinteri, Giuseppe Lacidogna, and Oscar Borla

6 Elemental Content Variations in Crushed Mortar Specimens Measured by Instrumental Neutron Activation Analysis (INAA) 73
Alberto Carpinteri, Oscar Borla, and Giuseppe Lacidogna

7 Piezonuclear Evidences from Tensile and Compression Tests on Steel 83
Stefano Invernizzi, Oscar Borla, Giuseppe Lacidogna, and Alberto Carpinteri

8   **Cold Nuclear Fusion Explained by Hydrogen Embrittlement and Piezonuclear Fissions in Metallic Electrodes: Part I: Ni-Fe and Co-Cr Electrodes** 99
    Alberto Carpinteri, Oscar Borla, Alessandro Goi, Amedeo Manuello, and Diego Veneziano

9   **Cold Nuclear Fusion Explained by Hydrogen Embrittlement and Piezonuclear Fissions in Metallic Electrodes: Part II: Pd and Ni Electrodes** 123
    Alberto Carpinteri, Oscar Borla, Alessandro Goi, Salvatore Guastella, Amedeo Manuello, and Diego Veneziano

10 **Piezonuclear Neutron Emissions from Earthquakes and Volcanic Eruptions** 135
    Oscar Borla, Giuseppe Lacidogna, and Alberto Carpinteri

11 **Is the Shroud of Turin in Relation to the Old Jerusalem Historical Earthquake?** 153
    Alberto Carpinteri, Giuseppe Lacidogna, and Oscar Borla

12 **Evolution and Fate of Chemical Elements in the Earth's Crust, Ocean, and Atmosphere** 163
    Alberto Carpinteri and Amedeo Manuello

13 **Chemical Evolution in the Earth's Mantle and Its Explanation Based on Piezonuclear Fission Reactions** 183
    Alberto Carpinteri, Amedeo Manuello, and Luca Negri

14 **Piezonuclear Fission Reactions Triggered by Fracture and Turbulence in the Rocky and Gaseous Planets of the Solar System** 197
    Alberto Carpinteri, Amedeo Manuello, and Luca Negri

15 **Piezonuclear Fission Reactions Simulated by the Lattice Model of the Atomic Nucleus** 219
    Norman D. Cook, Amedeo Manuello, Diego Veneziano, and Alberto Carpinteri

16 **Correlated Fracture Precursors in Rocks and Cement-Based Materials Under Stress** 237
    Gianni Niccolini, Oscar Borla, Giuseppe Lacidogna, and Alberto Carpinteri

17 **The Sacred Mountain of Varallo Renaissance Complex in Italy: Damage Analysis of Decorated Surfaces and Structural Supports** 249
    Federico Accornero, Stefano Invernizzi, Giuseppe Lacidogna, and Alberto Carpinteri

Chapter 1
TeraHertz Phonons and Piezonuclear Reactions from Nano-scale Mechanical Instabilities

Alberto Carpinteri

Abstract TeraHertz phonons are produced in condensed matter by mechanical instabilities at the nano-scale (fracture, turbulence, buckling). They present a frequency that is close to the resonance frequency of the atomic lattices and an energy that is close to that of thermal neutrons. A series of fracture experiments on natural rocks has recently demonstrated that the TeraHertz phonons are able to induce fission reactions on medium weight elements with neutron and/or alpha particle emissions. The same phenomenon appears to have occurred in several different situations and to explain puzzles related to the history of our planet, like the ocean formation or the primordial carbon pollution, as well as scientific mysteries, like the so-called "cold nuclear fusion" or the correct radio-carbon dating of organic materials.

Very important applications to earthquake precursors, climate change, energy production, and cell biology can not be excluded.

Keywords Fracture • Turbulence • Buckling • TeraHertz phonons • Ultrasonic pressure waves • Piezonuclear fission reactions • Neutron emissions • Alpha particle emissions • Acoustic emissions • Electromagnetic emissions • Compositional changes • Great Oxidation Event • Carbon pollution • Earthquake precursors • Chemical evolution • Great Red Spot of Jupiter • Cold nuclear fusion • Turin Shroud • Cell mechanotransduction • Protein folding

From the Distinguished Lecture in Solid Mechanics presented by Professor Alberto Carpinteri at the California Institute of Technology on May 30, 2014.

A. Carpinteri (✉)
Department of Structural, Geotechnical and Building Engineering, Politecnico di Torino, Corso Duca degli Abruzzi 24, 10129 Torino, Italy
e-mail: alberto.carpinteri@polito.it

1.1 Fracture and Acoustic Emission: From Hertz to TeraHertz Pressure Wave Frequencies

When you cut a stretched rubber band, it remains subject to rapid fluctuations for a few moments. The same phenomenon occurs in any solid body when it breaks in a brittle way, even if only partially. In the case of the formation or propagation of micro-cracks, such dynamic phenomenon appears under the form of longitudinal waves of expansion/contraction (tension/compression), in addition to transverse or shear waves. These are generally said pressure waves, or phonons when their particle nature is emphasized, and travel at a speed which is characteristic of the medium, and, for most of the solids and fluids, presents an order of magnitude of 10^3 metre/second. On the other hand, the wavelength of pressure waves emitted by forming or propagating cracks appears to be of the same order of magnitude of crack size or crack advancement length. The wavelength can not therefore exceed the maximum size of the body in which the crack is contained and may vary from the nanometre scale (10^{-9} metres), for defects in crystal lattices such as vacancies and dislocations, up to the kilometre, in the case of Earth's Crust faults. Applying the well-known relationship: frequency = speed/wavelength, one obtains the two extreme cases corresponding to the frequency of pressure waves: 10^{12} oscillations/ second (TeraHertz), in the case of the formation of nano-cracks, as well as one oscillation/second (Hertz), in the case of large-scale tectonic dynamics (Fig. 1.1).

In fact in solids, whatever their size, the cracks that are formed or propagate are of different lengths, sometimes belonging to different orders of magnitude. In particular, in the nanoscopic bodies the only frequency present should be the TeraHertz, since the higher frequencies would imply defects at the atomic or subatomic scales. On the contrary, in the Earth's Crust and during an earthquake,

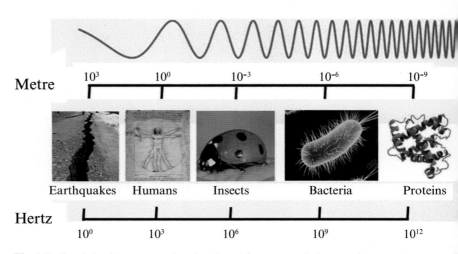

Fig. 1.1 Correlation between wavelength scale and frequency scale by assuming a constant pressure wave speed

cracking is a multi-scale phenomenon as well as the frequencies of pressure waves are spread over a broad spectrum (Fig. 1.1). Moreover, while at the earlier stages of the seismic event mainly small cracks will be present and active and therefore high frequencies, so at the end large cracks and low frequencies will prevail, the latter typically in the audible field.

1.2 Seismic Precursors: Acoustic, Electromagnetic, Neutron Emissions

Further considering the very important case of earthquakes, you can complete the picture by stating that, as fracture at the nanoscale (10^{-9} metres) emits phonons at the frequency scale of TeraHertz (10^{12} Hertz), so fracture at the microscale (10^{-6} metres) emits phonons at the frequency scale of GigaHertz (10^9 Hertz), at the scale of millimetre emits phonons at the scale of MegaHertz (10^6 Hertz), at the scale of metre emits phonons at the scale of kiloHertz (10^3 Hertz), and eventually faults at the kilometre scale emit phonons at the scale of the simple Hertz, which is the typical and most likely frequency of seismic oscillations (Fig. 1.1) [1].

The animals with sensitive hearing in the ultrasonic field (frequency > 20 kilo-Hertz) "feel" the earthquake up to one day in advance, when the active cracks are still below the metre scale. Ultrasounds are in fact a well-known seismic precursor [2, 3]. With frequencies between Mega- and GigaHertz, and therefore cracks between the micron and the millimetre scale, phonons can generate electromagnetic waves of the same frequency, which turn out to be even a more advanced seismic precursor (up to a few days before) [4, 5].

When phonons show frequencies between Giga- and TeraHertz, and then with cracks below the micron scale, we are witnessing a phenomenon partially unexpected: phonons resonate with the crystal lattices and, through a complex cascade of events (acceleration of electrons, bremsstrahlung gamma radiation, photofission, etc.), may produce nuclear fission reactions [6–14]. It can be shown experimentally how such fission reactions can emit neutrons [15–17] like in the well-known case of uranium-235 but without gamma radiation and radioactive wastes. Note that the Debye frequency, i.e., the fundamental frequency of free vibration of crystal lattices, is around the TeraHertz, and this is not a coincidence, since it is simply due to the fact that the inter-atomic distance is just around the nanometre, as indeed the minimum size of the lattice defects. As the chain reactions are sustained by thermal neutrons in a nuclear power plant, so the piezonuclear reactions are triggered by phonons that have a frequency close to the resonance frequency of the crystal lattice and an energy close to that of thermal neutrons. Neutrons therefore appear to be as the most advanced earthquake precursor (up to three weeks before) [18–23].

1.3 Chemical Evolution of Our Planet and Its Reproduction in the Fracture Mechanics Laboratory

The piezonuclear fission reactions appear then to be induced by pressure waves at very high frequencies (TeraHertz). They are often accompanied and revealed by the emission of neutrons and/or alpha particles. However, gamma rays and radioactive wastes appear to be absent in the experiments. Ultrasonic pressure waves may in turn be produced by the most common mechanical instabilities, such as fracture in solids and turbulence in fluids. Both are hierarchical, multi-fractal, and dissipative phenomena, where cracks and vortexes, respectively, are present at the different scales.

After the earlier experiments conducted at the National Research Council of Italy (CNR) [24, 25], soliciting with ultrasounds aqueous solutions of iron salts, the research group of the Politecnico di Torino has conducted fracture experiments on solid samples, using iron-rich rocks like granite [26–36], basalt and magnetite [37, 38], and then marble [39], mortar [40], and steel [41]. Different types of detectors have demonstrated the presence of significant neutron emissions, in some cases by different orders of magnitude higher than the usual environmental background (up to 10 times from granitic rocks, up to 100 times from basalt, up to 1000 times from magnetite).

The neutron flux was found to depend, besides on the iron content, on the size of the specimen through the well-known brittleness size effect [42–45]: larger sizes imply a higher brittleness, i.e. a more relevant strain energy release, and therefore more neutrons.

These studies have also been able to give an answer to some puzzles related to the history of our planet. It has been shown how the piezonuclear reactions that would have occurred between 3.8 and 2.5 billion years ago, during the period of formation and most intense activity of tectonic plates, have resulted in the splitting of atoms of certain elements, which were so transformed into other lighter ones. Since the product-elements, i.e., the fragments of the fissions, appear to be stable isotopes, all the excess neutrons are therefore emitted. Several of the most abundant chemical elements have been involved in similar transformations, like a part of magnesium that transformed into carbon, forming the dense atmospheres of carbon of the primordial terrestrial eras [46, 47]. In a similar way, calcium depletion contributed to the formation of oceans as a result of fracture phenomena in limestone rocks.

Considering the entire life of our planet and all the most abundant chemical elements [48–50], it can be seen how ferrous elements have dramatically decreased in the Earth's Crust (-12 %), as well as at the same time aluminum and silicon have increased (+8.8 %). An increment in magnesium (+3.2 %), which then transformed into carbon, has been assumed as the origin of carbon-rich primordial atmospheres. Similarly, alkaline-earth elements have strongly decreased (-8.7 %), whereas alkaline elements (+5.4 %) and oxygen (+3.3 %) have increased. The appearance of a 3.3 % oxygen represents the well-known Great Oxidation Event, a phenomenon that led to the formation of oceans and the origin of life on our planet.

These transformations, that have lasted for billion years in the Earth's Crust, have been reproduced in the laboratory in a fraction of a second by crushing different rock samples. We were able to confirm, through advanced micro-chemical analyses, the most relevant compositional variations described above at the geological and planetary scales: the transformation of iron into aluminum, or into magnesium and silicon (in iron-rich natural rocks [37]), as well as the transformation of calcium and magnesium into other lighter elements including carbon (in the samples of marble [39]). Such variations are shown to be not modest at all. The iron decrement in magnetite was found to be of 27.9 %, compared to an overall increment of 27.7 % in lighter elements. So in marble, carbon has increased by 13 %, compared to an exactly equivalent overall decrement in heavier elements.

Since the natural carbon production of the primordial eras, although at a slower rate, is going on even today, due to the seismic activity, the monitoring of carbon dioxide In relation to major earthquakes can be considered as a potential earthquake precursor [47], in addition to acoustic, electromagnetic, and neutron emissions.

1.4   Chemical Evolution in the Solar System

Even in the case of the other planets of the Solar System we are witnessing a series of experimental evidences that can be interpreted in the light of piezonuclear fission reactions [51]. In particular, the data coming from different surveys on the crust of planet Mars, made available by the NASA space missions over the past 15 years, suggest that the increase in certain elements (iron, chlorine, and argon) and the concomitant decrease in others (nickel and potassium), together with the emission of neutrons from the major fault lines in the planet, should all be considered as phenomena directly correlated. These data provide a clear confirmation that seismic activity has contributed to the chemical evolution of the Red Planet. Similar experimental evidences are concerning Mercury, Jupiter (with its relevant emission of neutrons from the Great Red Spot), and the Sun itself. The piezonuclear phenomena are triggered by earthquakes in rocky planets and by storms in gaseous planets. In the Sun, for example, the drastic decrease in lithium appears to be due to the fission of the same lithium into helium and hydrogen.

1.5   A Plausible Explanation of the So-Called Cold Nuclear Fusion

Several evidences have been observed during the last 25 years of anomalous nuclear reactions in electrolytic experiments. The purpose was that of providing an explanation to the phenomena related to the so-called "cold nuclear fusion", and of evaluating the possibility of a heat generation from electrolytic cells [52–54]. Despite the large amount of positive experimental results, the understanding of these phenomena is still unsatisfactory. On the other hand, as reported in most of the articles on cold nuclear fusion, the appearance of micro-cracks on the surface of the electrodes used in the experiments is one of the most common observations. It is therefore possible to give an explanation of a mechanical nature which takes into account the hydrogen embrittlement of the metallic electrodes.

In our earlier experiments [55, 56], electrolytic phenomena were produced by means of an anode of a nickel-iron alloy and a cathode of a cobalt-chromium alloy, immersed in an aqueous solution of potassium carbonate. During these experiments, emissions of neutrons and alpha particles were revealed. Furthermore, the composition of the electrodes was analyzed before and after the experiments, allowing to identify piezonuclear fissions occurred in the electrodes. The primary process appears to be a symmetric fission of the atom of nickel into two atoms of silicon, or two atoms of magnesium. In the latter case, additional fragments were found to be constituted of alpha particles.

In our later experiments [57, 58], where a palladium electrode was used, the primary process appears to be the non-symmetric fission of palladium into iron and calcium, whereas the secondary processes appear to be the further fissions of both such products into oxygen atoms and alpha particles.

1.6 A Catastrophic Earthquake Behind the Mystery of the Shroud

A neutron radiation, produced by the historic earthquake occurred around the year A.D. 33, may have caused the erroneous radiocarbon dating of the Shroud of Turin in 1988 [59, 60]. Neutron radiation could have also caused the image formation of a crucified man, who many believe to be Jesus Christ, on the linen cloth. The Shroud has attracted a large interest since Secondo Pia took the first photograph in 1898: in fact, one wonders if this is really the shroud of Jesus Christ, investigating its true age and the manner in which the image was produced. According to carbon-14 dating, the cloth would approximately be only 750 years old. From 1988 onwards, several researchers have instead argued that the Shroud would be much older and that the process of dating would have been wrong because of neutron radiation, so as to form new isotopes of carbon from nitrogen atoms. However, so far, any plausible physical reason has not been yet identified that can justify the origin of this neutron radiation [61]. The mechanical and chemical experiments described in the previous sections allow, on the other hand, to hypothesize that high-frequency pressure waves, generated in the Earth's Crust by the historical earthquake of AD 33, which took place in old Jerusalem with a magnitude between 8 and 9 on the Richter Scale, may have produced a neutron radiography on linen fibers and seemingly rejuvenated the same fabric. Let us consider that, although the calculated integral flux of 10^{13} neutrons per square centimetre is 10 times greater than the cancer therapy dose, nevertheless it is 100 times smaller than the lethal dose.

1.7 Future Applications Also to Biology?

Regarding the living cells, the piezonuclear reactions could explain the mechanism that governs the so-called "sodium-potassium pump" and, more in general, the metabolic processes. In the case of the ionic pump, the ions of sodium and potassium would be subject to a continuous and recurrent transformation of one into the other, losing and regaining an oxygen atom at each passage through the cell membrane. As cells are microscopic objects, so proteins are nanoscopic and the typical mechanisms of "folding", which make the passage of ions through the cell membrane possible, are accompanied by vibrational phenomena of resonance at the frequency of TeraHertz [62, 63] (mechanotransduction [64]). More precisely, the folding changes of configuration in the proteins should be interpreted as a dynamic nano-buckling (with snap-through) of thin complex-shaped shells. As in the case of acoustic emission from cracks or vortexes at the nano-scale, also the resonance frequency of nano-structures may be evaluated in the TeraHertz range.

Analogous reasons could explain also the "digestion" of radioactive isotopes intended as their transformation into stable isotopes of chemical elements which are essentials for the vital activity of microbial cultures [65].

References

1. Ashcroft NW, Mermin DN (2013) Solid state physics. Cengage Learning, Delhi
2. Lockner DA et al (1991) Quasi-static fault growth and shear fracture energy in granite. Nature 350:39–42
3. Carpinteri A, Lacidogna G, Niccolini G (2007) Acoustic emission monitoring of medieval towers considered as sensitive earthquake receptors. Nat Hazards Earth Syst Sci 7:251–261
4. Rabinovitch A, Frid V, Bahat D (2007) Surface oscillations. A possible source of fracture induced electromagnetic oscillations. Tectonophysics 431:15–21
5. Carpinteri A et al (2012) Mechanical and electromagnetic emissions related to stress-induced cracks. Exp Tech 36:53–64
6. Bridgman PW (1927) The breakdown of atoms at high pressures. Phys Rev 29:188–191
7. Batzel RE, Seaborg GT (1951) Fission of medium weight elements. Phys Rev 82:607–615
8. Fulmer CB et al (1967) Evidence for photofission of iron. Phys Rev Lett 19:522–523
9. Widom A, Swain J, Srivastava YN (2013) Neutron production from the fracture of piezoelectric rocks. J Phys G: Nucl Part Phys 40(15006):1–8
10. Widom A, Swain J, Srivastava YN (2015) Photo-disintegration of the iron nucleus in fractured magnetite rocks with magnetostriction. Meccanica 50:1205–1216
11. Hagelstein PL, Letts D, Cravens D (2010) Teraherz difference frequency response of PdD in two-laser experiments. J Cond Mat Nucl Sci 3:59
12. Hagelstein PL, Chaudhary IU (2015) Anomalies in fracture experiments and energy exchange between vibrations and nuclei. Meccanica 50:1189–1203
13. Cook ND (2010) Models of the atomic nucleus. Springer, Dordrecht
14. Cook ND, Manuello A, Veneziano D, Carpinteri A (2016) Piezonuclear fission reactions simulated by the lattice model of the atomic nucleus. In: Carpinteri A, Lacidogna G, Manuello A (eds) Acoustic, electromagnetic, neutron emissions from fracture and earthquakes. Springer, Cham

15. Diebner K (1962) Fusionsprozesse mit hilfe konvergenter stosswellen – einige aeltere und neuere versuche und ueberlegungen. Kerntechnik 3:89–93
16. Derjaguin BV et al (1989) Titanium fracture yields neutrons? Nature 34:492
17. Fujii MF et al (2002) Neutron emission from fracture of piezoelectric materials in deuterium atmosphere. Jpn J Appl Phys Pt1 41:2115–2119
18. Sobolev GA, Shestopalov IP, Kharin EP (1998) Implications of solar flares for the seismic activity of the Earth, Izvestiya, Phys. Solid Earth 34:603–607
19. Volodichev NN et al. (1999) Lunar periodicity of the neutron radiation burst and seismic activity on the Earth. Proceedings of the 26th International Cosmic Ray Conference, Salt Lake City, 17–25 Aug 1999
20. Kuzhevskij M, Nechaev OY, Sigaeva EA (2003) Distribution of neutrons near the Earth's surface. Nat Hazards Earth Syst Sci 3:255–262
21. Kuzhevskij M (2003) Neutron flux variations near the Earth's crust. A possible tectonic activity detection. Nat Hazards Earth Syst Sci 3:637–645
22. Sigaeva EA et al (2006) Thermal neutrons' observations before the Sumatra earthquake. Geophys Res Abstr 8:00435
23. Borla O, Lacidogna G, Carpinteri A (2016) Piezonuclear neutron emissions from earthquakes and volcanic eruptions. In: Carpinteri A, Lacidogna G, Manuello A (eds) Acoustic, electromagnetic, neutron emissions from fracture and earthquakes. Springer, Cham
24. Cardone F, Mignani R (2007) Deformed spacetime, Chapters 16 and 17. Springer, Dordrecht
25. Cardone F, Cherubini G, Petrucci A (2009) Piezonuclear neutrons. Phys Lett A 373:862–866
26. Carpinteri A, Cardone F, Lacidogna G (2010) Energy emissions from failure phenomena: mechanical, electromagnetic, nuclear. Exp Mech 50:1235–1243
27. Carpinteri A, Lacidogna G, Manuello A, Borla O (2013) Piezonuclear fission reactions from earthquakes and brittle rocks failure: evidence of neutron emission and nonradioactive product elements. Exp Mech 53(3):345–365
28. Carpinteri A, Cardone F, Lacidogna G (2009) Piezonuclear neutrons from brittle fracture: early results of mechanical compression tests. Strain 45:332–339
29. Carpinteri A, Chiodoni A, Manuello A, Sandrone R (2011) Compositional and microchemical evidence of piezonuclear fission reactions in rock specimens subjected to compression tests. Strain 47(2):267–281
30. Carpinteri A, Manuello A (2011) Geomechanical and Geochemical evidence of piezonuclear fission reactions in the Earth's crust. Strain 47(2):282–292
31. Carpinteri A, Borla O, Lacidogna G, Manuello A (2010) Neutron emissions in brittle rocks during compression tests: monotonic vs. cyclic loading. Phys Mesomech 13:264–274
32. Carpinteri A, Manuello A (2012) An indirect evidence of piezonuclear fission reactions: geomechanical and geochemical evolution in the Earth's crust. Phys Mesomech 15:14–23
33. Cardone F, Carpinteri A, Lacidogna G (2009) Piezonuclear neutrons from fracturing of inert solids. Phys Lett A 373:4158–4163
34. Carpinteri A, Lacidogna G, Manuello A, Borla O (2011) Energy emissions from brittle fracture: neutron measurements and geological evidences of piezonuclear reactions. Strength, Fracture, Complexity 7:13–31
35. Carpinteri A, Lacidogna G, Manuello A, Borla O (2012) Piezonuclear fission reactions in rocks: evidences from microchemical analysis, neutron emission, and geological transformation. Rock Mech Rock Eng 45(4):445–459
36. Carpinteri A, Lacidogna G, Borla O, Manuello A, Niccolini G (2012) Electromagnetic and neutron emissions from brittle rocks failure: experimental evidence and geological implications. Sadhana 37(1):59–78
37. Manuello A et al (2016) Neutron emissions and compositional changes at the compression failure of iron-rich natural rocks. In: Carpinteri A, Lacidogna G, Manuello A (eds) Acoustic, electromagnetic, neutron emissions from fracture and earthquakes. Springer, Cham
38. Carpinteri A et al (2016) Frequency-dependent neutron emissions during fatigue tests on iron-rich natural rocks. In: Carpinteri A, Lacidogna G, Manuello A (eds) Acoustic, electromagnetic, neutron emissions from fracture and earthquakes. Springer, Cham

39. Carpinteri A, Lacidogna G, Borla O (2016) Alpha particle emissions from Carrara marble specimens crushed in compression and X-ray Photoelectron Spectroscopy of correlated nuclear transmutations. In: Carpinteri A, Lacidogna G, Manuello A (eds) Acoustic, electromagnetic, neutron emissions from fracture and earthquakes. Springer, Cham
40. Carpinteri A, Borla O, Lacidogna G (2016) Elemental content variations in crushed mortar specimens measured by Instrumental Neutron Activation Analysis (INAA). In: Carpinteri A, Lacidogna G, Manuello A (eds) Acoustic, electromagnetic, neutron emissions from fracture and earthquakes. Springer, Cham
41. Invernizzi S et al (2016) Piezonuclear evidences from tensile and compression tests on steel. In: Carpinteri A, Lacidogna G, Manuello A (eds) Acoustic, electromagnetic, neutron emissions from fracture and earthquakes. Springer, Cham
42. Hudson JA, Crouch SL, Fairhurst C (1972) Soft, stiff and servo-controlled testing machines: a review with reference to rock failure. Eng Geol 6:155–189
43. Carpinteri A (1989) Cusp catastrophe interpretation of fracture instability. J Mech Phys Solids 37:567–582
44. Carpinteri A, Pugno N (2005) Are scaling laws of strength of solids related to mechanics or to geometry? Nat Mater 4:421–423
45. Carpinteri A, Corrado M (2009) An extended (fractal) overlapping crack model to describe crushing size-scale effects in compression, Eng. Fail Anal 16:2530–2540
46. Liu L (2004) The inception of the oceans and CO2-atmosphere in the early history of the Earth, Earth Planet. Sci Lett 227:179–184
47. Padron E et al (2008) Changes on diffuse CO2 emission and relation to seismic activity in and around El Hierro. Canary Islands. Pure Appl Geophys 165:95–114
48. Taylor SR, McLennan SM (2009) Planetary crusts: their composition, origin and evolution. Cambridge University Press, Cambridge
49. Carpinteri A, Manuello A (2016) Evolution and fate of chemical elements in the Earth's crust, ocean, and atmosphere. In: Carpinteri A, Lacidogna G, Manuello A (eds) Acoustic, electromagnetic, neutron emissions from fracture and earthquakes. Springer, Cham
50. Carpinteri A, Manuello A, Negri L (2016) Chemical evolution in the Earth's mantle and its explanation based on piezonuclear fission reactions. In: Carpinteri A, Lacidogna G, Manuello A (eds) Acoustic, electromagnetic, neutron emissions from fracture and earthquakes. Springer, Cham
51. Carpinteri A, Manuello A, Negri L (2016) Piezonuclear fission reactions triggered by fracture and turbulence in the rocky and gaseous planets of the Solar System. In: Carpinteri A, Lacidogna G, Manuello A (eds) Acoustic, electromagnetic, neutron emissions from fracture and earthquakes. Springer, Cham
52. Fleischmann M, Pons S (1989) Electrochemically induced nuclear fusion of deuterium. J Electroanal Chem 261:301–308
53. Preparata G (1991) A new look at solid-state fractures, particle emissions and «cold» nuclear fusion. Il Nuovo Cimento 104 A:1259–1263
54. Mizuno T (1998) The reality of cold fusion. Cold Fusion Technology, Concord
55. Carpinteri A et al (2013) Mechanical conjectures explaining cold nuclear fusion. Conf Exposition Exp Appl Mech (SEM), Lombard, Illinois 3:353–367
56. Carpinteri A et al (2016) Cold nuclear fusion explained by hydrogen embrittlement and piezonuclear fissions at the metallic electrodes – Part I: Ni-Fe and Co-Cr electrodes. In: Carpinteri A, Lacidogna G, Manuello A (eds) Acoustic, Electromagnetic, Neutron Emissions from Fracture and Earthquakes. Springer, Cham
57. Carpinteri A et al (2014) Hydrogen embrittlement and "cold fusion" effects in palladium during electrolysis experiments. Conf Exposition Exp Appl Mech (SEM), Greenville, South Carolina 6:37–47
58. Carpinteri A et al (2016) Cold nuclear fusion explained by hydrogen embrittlement and piezonuclear fissions at the metallic electrodes – Part II: Pd and Ni electrodes. In: Carpinteri A, Lacidogna G, Manuello A (eds) Acoustic, electromagnetic, neutron emissions from fracture and earthquakes. Springer, Cham

59. Carpinteri A, Lacidogna G, Manuello A, Borla O (2012) Piezonuclear neutrons from earth quakes as a hypothesis for the image formation and the radiocarbon dating of the Turin Shroud Sci Res Essays 7(29):2603–2612
60. Carpinteri A, Lacidogna G, Borla O (2016) Is the Shroud of Turin in relation to the Old Jerusalem historical earthquake? In: Carpinteri A, Lacidogna G, Manuello A (eds) Acoustic electromagnetic, neutron emissions from fracture and earthquakes. Springer, Cham
61. Phillips TJ, Hedges REM (1989) Shroud irradiated with neutrons? Nature 337:594
62. Acbas G et al (2014) Optical measurements of long-range protein vibrations. Nature Comm doi:10.1038/ncomms4076
63. http://www.buffalo.edu/news/releases/2014/01/012.html
64. Mofard MRK, Kamm RD (2010) Cellular mechanotransduction. Cambridge University Press Cambridge
65. Vysotskii VI, Kornilova AA (2010) Nuclear transmutation of stable and radioactive isotopes in biological systems. Pentagon Press, New Delhi

Chapter 2
Correlation Between Acoustic and Other Forms of Energy Emissions from Fracture Phenomena

Giuseppe Lacidogna, Oscar Borla, Gianni Niccolini, and Alberto Carpinteri

Abstract In the present investigation, acoustic (AE), electromagnetic (EME), and neutron (NE) emissions were measured during laboratory compression tests on rock specimens loaded up to failure. All the signals were acquired by a National Instruments Digitizer with eight channels simultaneously sampling. The aim was to find a time correlation between these three different forms of energy emission from rocks under compression. The tests were performed on magnetite and basalt specimens at constant displacement rate. AE signals were detected by applying to the specimen surface a piezoelectric (PZT) transducer with resonance frequency of about 150 kHz. EM signals were revealed by the current induced in a closed circuit due to change in the magnetic flux during specimen compression. The specimens were also monitored by means of a He^3 proportional neutron detector. During the tests were first detected the AE signals, and then the EM emission. All the recorded signals were correlated to the load vs time diagrams. The EM signals were obtained, in particular, during the typical snap-back instabilities, which characterize the load versus displacement diagrams of brittle materials such as rocks in compression. Neutron emission signals were generally identified at the end of the tests. As a matter of fact, neutron bursts usually occur when the behaviour of the specimen in compression is particularly brittle. Applications of these monitoring techniques to earthquake forecasting seem to be possible.

Keywords Compression • Brittle failure • Acoustic emission • Electromagnetic emission • Neutron measurements • Piezonuclear reactions

G. Lacidogna (✉) • O. Borla • G. Niccolini • A. Carpinteri
Department of Structural, Geotechnical and Building Engineering, Politecnico di Torino, Corso Duca degli Abruzzi 24, 10129 Torino, Italy
e-mail: giuseppe.lacidogna@polito.it

© Springer International Publishing Switzerland 2015   11
A. Carpinteri et al. (eds.), *Acoustic, Electromagnetic, Neutron Emissions from Fracture and Earthquakes*, DOI 10.1007/978-3-319-16955-2_2

2.1 Introduction

It is possible to demonstrate experimentally that the failure phenomena, in particular when they occur in a brittle way, i.e. with a mechanical energy release, emit additional forms of energy related to the fundamental natural forces. The authors have found experimental evidence and confirmation that energy emission of different forms occurs from solid-state fractures. The tests were carried out at the Laboratory of Fracture Mechanics of the Politecnico di Torino, Italy. By subjecting quasi-brittle materials such as granitic rocks to compression tests, for the first time, bursts of neutron emission (NE) during the failure process were observed [1–5], necessarily involving nuclear reactions, besides the well-known acoustic emission (AE) [6–13], and the phenomenon of electromagnetic radiation (EM) [14–19], which is highly suggestive of charge redistribution during material failure and at present under investigation.

The phenomenon of EME is regarded as an important precursor of critical phenomena in Geophysics, such as rock fractures, volcanic eruptions, and earthquakes [19, 20]. For example, anomalous radiations of geo-electromagnetic waves were observed before major earthquakes. At the laboratory scale, rocks and concrete under compression generate AE and EM emissions nearly simultaneously. This evidence suggests that also NE emissions are generated during crack growth, reinforcing the idea that the NE phenomenon can be applied as a forecasting tool for earthquakes.

While the mechanism of AE is fully understood, being provided by transient elastic waves due to stress redistribution following fracture propagation [6–13], the origin of EME from fracture is not completely clear and different attempts have been made to explain it.

An explanation of the EME origin was related to dislocation phenomena [16], which however are not able to explain EME from fracture in brittle materials, where the motion of dislocations can be neglected [17]. Frid et al. [17] and Rabinovitch et al. [21] recently proposed a model of the EME origin where, following the rupture of bonds during the crack growth, mechanical and electrical equilibrium are broken at the fracture surfaces with the creation of ions moving collectively as a surface wave on both faces. Lines of positive ions on both newly created faces (which maintain their charge neutrality unlike the capacitor model) oscillate collectively around their equilibrium positions in opposite phase to the negative ones. The resulting oscillating dipoles created on both faces of the propagating fracture act as the source of EME.

As regards the neutron emissions, in this paper we present experimental tests performed on brittle rocks (Magnetite and Basalt), using a He^3 neutron device and a bubble type BD thermodynamic neutron detectors. For brittle specimens of sufficiently large dimensions, neutron emissions, detected by He^3, were found to be up to three orders of magnitude higher than the ordinary natural background level at the time of the catastrophic failure. These emissions fully confirm the previous tests [1–5] and are due to piezonuclear reactions, which depend on the different modalities of energy release during the tests. For specimens with sufficiently large size and slenderness, a relatively high energy release is expected, and hence a higher probability of neutron emissions at the time of failure.

The experimental analysis carried out by the authors may open a new possible scenario, in which the stress state of the elements firstly involves the generation of microcracks, accompanied by mechanical energy release in the field of ultrasonic vibrations that can be measured using suitable AE equipments. Hence, the formation of coherent EM fields occurs over a wide range of frequencies, from few Hz to MHz, and even up to microwave frequencies. This excited state of the matter could be a cause of subsequent resonance phenomena of nuclei able to produce neutron bursts. This hypothesis is also confirmed by Widom et al. [22, 23]. As a matter of fact, the microcracking elastic energy release ultimately yields the acoustic vibrations, which are converted into electromagnetic oscillations. The electromagnetic waves, generated during microcracking, accelerate the condensed matter electrons which then collide with protons producing neutrons and neutrinos [22, 23].

2.2 Experimental Set Up

Experimental compression tests were performed on brittle rock specimens under monotonic displacement control. The materials used for the tests are non-radioactive Magnetite and Basalt. In these tests, a total of 29 cylindrical specimens with different size and slenderness are used (Fig. 2.1). The compression tests were performed at the Fracture Mechanics Laboratory of the Politecnico di Torino. In Table 2.1, the experimental data concerning the tested specimens are summarized.

All the specimens were subjected to uniaxial compression using a MTS servo-controlled hydraulic testing machine with a maximum capacity of 1000 kN. Each test was performed in piston travel displacement control by setting constant piston velocity. The specimens were arranged in contact with the press platens without any coupling material, according to the testing modalities known as "test by means of rigid platens with friction".

Fig. 2.1 Magnetite (*left*) and Basalt (*right*) cylindrical specimens, by varying slenderness and size-scale

Table 2.1 Tested specimens and their mechanical characteristics

| Specimens | Specimens number | Dimension | | Piston velocity | Volume | Average peak load |
		Diameter [mm]	Slenderness λ	[m/s]	[mm^3]	[kN]
Magnetite						
M-20-0.5	5	20	0.5	5×10^{-7}	3˙140	67.46
M-20-1	2	20	1	5×10^{-7}	6˙280	48.20
M-20-2	4	20	2	5×10^{-7}	12˙560	45.88
M-40-0.5	2	40	0.5	1×10^{-6}	25˙120	159.25
M-40-1	6	40	1	1×10^{-6}	50˙240	146.87
M-40-2	4	40	2	1×10^{-6}	100˙480	109.40
M-90-1	4	90	1	2×10^{-6}	572˙265	849.89
Basalt						
B-50-2	2	50	2	1×10^{-6}	196˙250	177.64

| Specimens | AE | | EME | | NE | |
	Average frequency [Hz]	Average highest Frequency [Hz]	Average frequency [Hz]	Average highest frequency [Hz]	Average neutron background [10^{-2} cps]	Average count rate at the neutron emission [10^{-2} cps]
Magnetite						
M-20-0.5	20˙975	56˙990	19˙622	>46˙000	4.84 ± 1.21	Background
M-20-1	25˙889	49˙488	—	—	5.90 ± 1.48	Background
M-20-2	23˙569	42˙984	32˙312	>55˙000	5.70 ± 1.42	Background
M-40-0.5	53˙354	133˙133	28˙278	>49˙000	5.50 ± 1.38	Background
M-40-1	60˙055	90˙997	37˙787	>77˙000	5.06 ± 1.26	18.75 ± 4.69
M-40-2	33˙520	73˙385	35˙274	>98˙000	5.60 ± 1.40	25.96 ± 6.49
M-90-1	52˙350	132˙062	110˙089	>1 MHz	4.95 ± 1.24	901.20 ± 225.30
Basalt						
B-50-2	76˙668	165˙363	204˙417	>335˙000	5.60 ± 1.40	14.22 ± 3.55

The AE activity emerging from the compressed specimens was detected by attaching to the specimen surface a piezoelectric (PZT) transducer, resonant at about 150 kHz, which is able to convert the high-frequency surface movements due to the acoustic wave into an electric signal (the AE signal). Sensitivity of the transducer in the low-frequency range was measured by placing it on a shaker excited with all frequencies in the range 0–10 kHz (white noise). The result of this calibration at low frequencies was 1.2 μV/(mm s^{-2}). Resonant sensors are more sensitive than broadband sensors, which are characterized by a flat frequency response in their working range, and then they can be successfully used in monitoring of large-sized structures.

The EME detecting device, realized at the National Research Institute of Metrology (INRIM), is constituted by three pickup coils with a different number of turns, made of a 0.2 mm copper wire, that are positioned around the monitored specimen. This instrumentation, which acquires data in the frequency range from few Hz up to 4 MHz, exploits the induction Faraday's law: the induced voltage in a closed circuit (loop) is proportional to the change in the magnetic flux throughout the circuit. The first coil, constituted by 5 turns, works in a frequency range from 300 kHz to 4 MHz. The other two coils constituted by 125 turns and 500 turns, work in the frequency range from few kHz to 20 kHz, and from few Hz to 1 kHz, respectively.

Due to the difficulties in neutron measurements in the presence of electromagnetic disturbances, EME measurement carries out both the validation of NE signals and the monitoring of EME from fracturing; the simultaneous presence of sharply peaked EME signals (well characterized and far from the continuous magnetic noise) and NE signals can be regarded as the signature of an ongoing damage process. As a further check on NE signals, a set of passive neutron detectors, based on the superheated bubble detection technique and insensitive to electromagnetic noise, were employed. A detailed description of the used neutron detectors is given.

2.3   He^3 Neutron Proportional Counter

The He^3 detector used in the compression tests under monotonic displacement control is a He^3 type (Xeram, France) with electronics of preamplification, amplification, and discrimination directly connected to the detector tube. The detector is powered with 1.3 kV, supplied via a high voltage NIM (Nuclear Instrument Module). The logic output producing the TTL (transistor-transistor logic) pulses is connected to a NIM counter. The device was calibrated for the measurement of thermal neutrons; its sensitivity is 65 cps/$n_{thermal}$ (± 10 % declared by the factory) i.e., a thermal neutron flux of 1 thermal neutron/s cm^2 corresponds to a count rate of 65 cps.

Considering that the fracture of dielectric materials, such as rocks, can lead to the emission of charged and neutral particles (electrons, photons, hard X-rays), in order to avoid possible false neutron measurements, the output of the detector is enabled for detecting signals only exceeding a fixed amplitude. This threshold value was determined by measuring the analog signal of the detector by means of a Co-60 gamma source (half-life: 5.271 years, type decay: beta$^-$, beta maximum energy: 317.8 keV, gammas: 1173.2 keV and 1332.5 keV). The presence of an interfering capacity on the charge preamplifier input increases the electronic noise and consequently the probability of spurious counts. For this reason, the coaxial cable used for connecting detector and charge preamplifier presented a low capacity (36 pF/m) and a short length (about 50 cm). Moreover, during the experimental measurements, the front-end electronics was screened with aluminum foils, and the He^3 tube was immersed in a sound-absorbing substance such as polystyrene in order to avoid possible accidental impacts and vibrations.

2.4   Neutron Bubble Detectors

A set of passive neutron detectors insensitive to electromagnetic noise and with zero gamma sensitivity was used in compression tests under cyclic loading. The dosimeters, based on superheated bubble detectors (BTI, Ontario, Canada) (Bubble Technology Industries (1992)) [24], are calibrated at the factory against an Am-Be source in terms of NCRP38 (NCRP report 38 (1971)) [25]. Bubble detectors provide instant visible detection and measurement of neutron dose. Each detector is composed of a polycarbonate vial filled with elastic tissue-equivalent polymer, in which droplets of a superheated gas (Freon) are dispersed. When a neutron strikes a droplet, the latter immediately vaporizes, forming a visible gas bubble trapped in the gel. The number of droplets provides a direct measurement of the equivalent neutron dose. These detectors are suitable for neutron integral dose measurements, in the energy ranges of thermal neutrons ($E = 0.025$ eV) and fast neutrons ($E > 100$ keV).

All the signals (AE, EME and NE) were acquired by a National Instruments Digitizer with eight channels simultaneously sampling at 1 MSa/s. The trigger was set to the AE channel with a detection threshold fixed at 20 mV to filter out the background noise.

For all the specimens, the recorded AE, EME and NE time series were related to the time history of the applied load.

2.5   Test Results

2.5.1   AE and EME Measurements

In this work, among all the 29 tested samples, the experimental results of four specimens (three of Magnetite and one of Basalt) are examined and described in detail. All specimens were tested in compression up to failure, showing a brittle response with a rapid decrease in load-carrying capacity when deformed beyond the peak load (Figs. 2.2 and 2.3). Experimental evidence indicates the presence of AE, EME and NE activity. In Figs. 2.2 and 2.3 the AE and EME bursts, i.e. the signals received by the devices, are reported as accumulated number and rate in time, while the NE bursts are reported as counts per second (cps). It is interesting to note that in this experimental campaign – unlike those carried out by the authors on other materials, such as concrete, Luserna stone, Carrara marble and Syracuse limestone [14, 15] – the EME activity is much more widespread during the loading process, and not just concentrated at the moment of the final collapse.

As a matter of fact, in the Magnetite specimen P1 (M-40-1 type), whose behaviour is described by the load vs. time curve in Fig. 2.2 (left), the observed bursts of AE and EME activity can be clearly correlated with the stress drops occurring before the collapse, Fig. 2.2 (upper and middle left). In the Magnetite

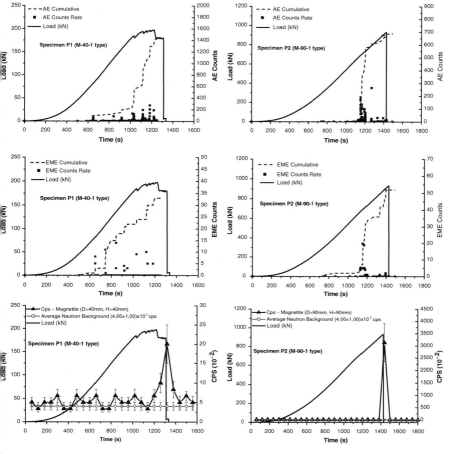

Fig. 2.2 Load vs. time diagram of the Magnetite specimen P1 (M-40-1 type) (*left*); accumulated number and rate of AEs (*upper left*); accumulated number and rate of EMEs (*middle left*); NE count rate (*lower left*). Load vs. time diagram of the Magnetite specimen P2 (M-90-1 type) (*right*); accumulated number and rate of AEs (*upper right*); accumulated number and rate of EMEs (*middle right*); NE count rate (*lower right*)

specimen P2 (M-90-1 type), characterized by a perfectly brittle behaviour without evident stress drops before the final collapse (note the linearity till failure of the load vs. time curve in Fig. 2.2 (right)), the specimen failure was preceded by two closely correlated bursts of AE and EME activity at nearly 80 % of the peak load, Fig. 2.2 (upper and middle right). This activity, particularly as regards electromagnetic emission, can be due to the behaviour under loading of Magnetite that – being rich in iron, about 65 % in weight – determines the formation of magnetic charges generated by friction during the loading process, and their spontaneous release independently of the formation of macro-cracks at the time of final collapse.

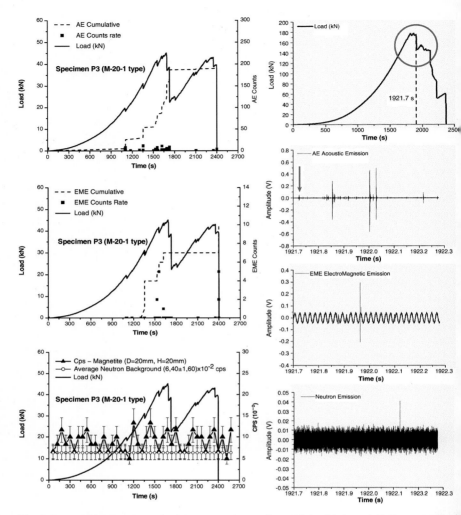

Fig. 2.3 Load vs. time diagram of the Magnetite specimen P3 (M-20-1 type) (*left*); accumulated number and rate of AEs (*upper left*); accumulated number and rate of EMEs (*middle left*); NE count rate (*lower left*). Load vs. time diagram of the Basalt specimen (*upper right*); AE, EME (*middle right*) and NE signals (*lower right*) detected in a time window of 0.6 s, starting at 1921.7 s from the beginning of the test

As a particular case, the load vs. time diagram of the Magnetite specimen P3 (M-20-1 type) is double-peaked with a significant stress drop at about 60 % of the test duration, followed by a drop in the AE rate, Fig. 2.3 (left). This momentary relaxation in the AE activity describes the well-known Kaiser effect [26], which states that, after stress drops, AE activity is very low during the reloading of the material until the stress exceeds the previous reached values, Fig. 2.3 (upper left)

This relaxation was observed also in the EME activity, Fig. 2.3 (middle left), confirming the close correlation degree between these two phenomena.

2.5.2   NE Measurements

As regards the NE measurements, the He^3 neutron detector was switched on at least one hour before the beginning of each compression test, in order to reach the thermal equilibrium of electronics, and to make sure that the behaviour of the device was stable with respect to intrinsic thermal effects. For the considered specimens P1, P2 and P3, the average measured background level ranges from $(4.00 \pm 1.00) \times 10^{-2}$ to $(6.40 \pm 1.60) \times 10^{-2}$ cps. In general, neutron measurements of specimens M-20-1 type yielded values comparable with the ordinary natural background, whereas in specimens M-40-1 type the experimental data exceeded the background value by about four times. For specimens M-90-1 type the neutron emissions achieved values up to three orders of magnitude higher than the ordinary background.

Moreover, the volumes of all the tested specimens are shown in Table 2.1. As reported by the Authors in a previous work [27], a volume approximately exceeding 200·000 mm^3, combined with the extreme brittleness of the tested material, represents a threshold value for a neutron emission of about one order of magnitude higher than the ordinary background.

In this case it is interesting to highlight as neutron measurements of specimens M-20-1 type (6·280 mm^3) yielded values comparable with the ordinary natural background, whereas in specimens M-40-1 type (50·240 mm^3) the experimental data exceeded the background value by about four times. For specimens M-90-1 type (572·265 mm^3) the neutron emissions achieved values up to three orders of magnitude higher than the ordinary background.

As regards the expected energy spectrum, it extends from thermal neutrons (0.025 eV) up to the fast component (few MeV). Also this behaviour has already been measured by the authors [1–5] by using specific devices such as proportional counters (He^3 devices) and passive bubble dosimeters. In the future, more detailed information about the energy spectrum using special spectrometers will be provided.

In Figs. 2.2 and 2.3 (left) the load vs. time diagram, and the neutron count rate evolution for each specimen are shown. Moreover, bursts of NE activity were observed at the failure time of specimens P1 (M-40-1 type) and P2 (M-90-1 type), Fig. 2.2 (lower), confirming the need of catastrophic ruptures, i.e., characterized by sudden release of the stored strain energy, to obtain such anomalous neutron emissions.

Furthermore, during the compression tests a rise in the thermal equivalent neutron dose, analysed by neutron bubble detectors, was measured, consistently with the increment in the neutron level measured by the He^3 device. In particular, for the specimen P2 (M-90-1 type), a value more than 1000 times higher than the ordinary background was found at the end of the test.

2.5.3 AE, EME and NE Time Correlation

Test results on the Basalt specimen P4 (B-50-2 type) give a high degree of correlation among the three emission time series and the load time history. As an example, considering a time window of 0.6 s starting at the 1921.7 s from the beginning of the test, AE bursts followed by an EME pulse in the kHz frequency gap are shown in Fig. 2.3 (right). Similar simultaneous EM pulses were observed in the Hz and MHz range that, for reasons of space, are not included in the same figure. The time window is related to the evident stress drop indicated by a circle in the load vs. time diagram, Fig. 2.3 (right).

As already discussed in [14], AE and EME signals, from a growing fracture, follow a time delay consistent with their propagation velocities. Being d the distance between source (fracture) and AE transducer, v_{AE} and v_{EME} the average propagation velocities of AE and EME waves with $v_{AE} << v_{EME}$, the time delay can be estimated by $\Delta t = d/v_{AE}$. Inserting $d = 10^{-1}$ m and $v_{AE} = 10^3$ m s^{-1}, an estimation of the time delay for the considered event is $\Delta t = 10^{-4}$ s $= 100$ µs.

If then we consider that the main crack propagation, in the specified time-window, takes place (begins its first motion, indicated with an arrow in Fig. 2.3 (right)) when the first AE peak of great amplitude is recorded, the main EME pulse follows the AE burst, although considering the different average propagation velocities of AE and EME signals. Therefore, the EME signal seems to spread during the mechanical vibration generated by fracture. Finally, the NE event is observed at the time of catastrophic failure of the specimen, Fig. 2.3 (upper and lower right).

Considering the behaviour of brittle specimens under mechanical loading, further interesting discussions on the variation in the AE vibration frequencies and on the signal peak distributions have also been reported [14, 15, 28, 29].

2.6 Conclusions

The experimental evidence presented in this paper confirms the previous investigations on AE and EME signals as collapse precursors in natural materials like rocks. The observed EME were strictly correlated in time with AE signals in all the tested specimens. Bursts of AE and EME activity were always observed when significant stress drops occur. This suggests the use of electromagnetic measurements to enhance monitoring systems based on the AE technique. In addition, NE activity was observed when the specimen fails in a sudden, catastrophic way. In particular, for specimens with sufficiently large size, the neutron flux was found to be about three orders of magnitude higher than the background level at the time of catastrophic failure. Therefore, the observed acoustic, electromagnetic, and neutron activity from laboratory experiments looks promising for effective applications also at the geophysical scale.

As a matter of fact — based on the analogy between AE and seismic activity [30–32], on the anomalous radiation of geoelectromagnetic waves observed before major earthquakes [33], and on recent experimental studies that measured neutron components exceeding the usual background in correspondence to seismic activity [34] — if we take into account the correlation between acoustic/electromagnetic/neutron emissions and seismic activity, it could be possible to set up a sort of alarm system based on a regional warning network.

The results obtained from this analysis show how the crack generation is accompanied by mechanical energy release in the field of ultrasonic vibrations detected by AE sensors. It was also observed that, for constant specimen diameter, the AE signals reached high frequency peaks for low slenderness values, whereas the EME frequencies increase with the sample size. The highest neutron emissions occurred from specimens with EME detected in the field of MHz. This shows that the formation of coherent EM fields (i.e. characterized by evident pulses generated by specific phase relationship between the electric field values at different times) occurs over a wide range of frequencies, from few Hz to MHz and even up to microwaves, during the fracture propagation. This excited state of the matter could be a cause of subsequent resonance phenomena of nuclei able to produce neutron bursts in the presence of stress-drops or sudden catastrophic fractures.

References

1. Carpinteri A, Cardone F, Lacidogna G (2009) Piezonuclear neutrons from brittle fracture: early results of mechanical compression tests. Strain 45:332–339
2. Cardone F, Carpinteri A, Lacidogna G (2009) Piezonuclear neutrons from fracturing of inert solids. Phys Lett A 373:4158–4163
3. Carpinteri A, Cardone F, Lacidogna G (2010) Energy emissions from failure phenomena: mechanical, electromagnetic, nuclear. Exp Mech 50:1235–1243
4. Carpinteri A, Borla O, Lacidogna G, Manuello A (2010) Neutron emissions in brittle rocks during compression tests: monotonic vs cyclic loading. Phys Mesomech 13:268–274
5. Carpinteri A, Lacidogna G, Manuello A, Borla O (2011) Energy emissions from brittle fracture: neutron measurements and geological evidences of piezonuclear reactions. Strenght Fract Complexity 7:13–31
6. Mogi K (1962) Study of elastic shocks caused by the fracture of heterogeneous materials and its relation to earthquake phenomena. Bull Earthquake Res Inst 40:125–173
7. Lockner DA, Byerlee JD, Kuksenko V, Ponomarev A, Sidorin A (1991) Quasi static fault growth and shear fracture energy in granite. Nature 350:39–42
8. Ohtsu M (1996) The history and development of acoustic emission in concrete engineering. Mag Concr Res 48:321–330
9. Rundle JB, Turcotte DL, Shcherbakov R, Klein W, Sammis C (2003) Statistical physics approach to understanding the multiscale dynamics of earthquake fault systems. Rev Geophys 41:1019–1049
10. Niccolini G, Schiavi A, Tarizzo P, Carpinteri A, Lacidogna G, Manuello A (2010) Scaling in temporal occurrence of quasi-rigid-body vibration pulses due to macrofractures. Phys Rev E 82:46115/1–46115/5
11. Carpinteri A, Lacidogna G (2006) Damage monitoring of an historical masonry building by the acoustic emission technique. Mater Struct 39:161–167

12. Carpinteri A, Lacidogna G (2006) Structural monitoring and integrity assessment of medieva towers. J Struct Eng (ASCE) 132:1681–1690
13. Carpinteri A, Lacidogna G (2007) Damage evaluation of three masonry towers by acousti emission. Eng Struct 29:1569–1579
14. Lacidogna G, Carpinteri A, Manuello A, Durin G, Schiavi A, Niccolini G, Agosto A (2010 Acoustic and electromagnetic emissions as precursor phenomena in failure processes. Strai 47(2):144–152
15. Carpinteri A, Lacidogna G, Manuello A, Niccolini A, Schiavi A, Agosto A (2010) Mechanica and electromagnetic emissions related to stress-induced cracks. Exp Tech 36(3):53–64
16. Misra A (1977) Theoretical study of the fracture-induced magnetic effect in ferromagneti materials. Phys Lett A 62:234–236
17. Frid V, Rabinovitch A, Bahat D (2003) Fracture induced electromagnetic radiation. J Phys L 36:1620–1628
18. Hadjicontis V, Mavromatou C, Nonos D (2004) Stress induced polarization currents an electromagnetic emission from rocks and ionic crystals, accompanying their deformation Nat Hazards Earth Syst Sci 4:633–639
19. Warwick JW, Stoker C, Meyer TR (1982) Radio emission associated with rock fracture Possible application to the great Chilean earthquake of May 22, 1960. J Geophys Re 87:2851–2859
20. Nagao T, Enomoto Y, Fujinawa Y et al (2002) Electromagnetic anomalies associated wit 1995 Kobe earthquake. J Geodynamics 33:401–411
21. Rabinovitch A, Frid V, Bahat D (2007) Surface oscillations. A possible source of fractur induced electromagnetic oscillations. Tectonophysics 431:15–21
22. Widom A, Swain J, Srivastava YN (2013) Neutron production from the fracture of piezoelec tric rocks. J Phys G Nucl Part Phys 40:15006 (1–8)
23. Widom A, Swain J, Srivastava YN (2015) Photo-disintegration of the iron nucleus in fracture magnetite rocks with magnetostriction. Meccanica 50:1205–1216
24. Bubble Technology Industries (1992) Instruction manual for the Bubble detector, Chalk River Ontario, Canada
25. National Council on Radiation Protection and Measurements (1971) Protection against Neu tron Radiation, NCRP Report 38
26. Kaiser J (1950) Ph. D. dissertation, Munich (FRG), Technische Hochschule München
27. Carpinteri A, Lacidogna G, Manuello A, Borla O (2013) Piezonuclear fission reactions fron earthquakes and brittle rocks failure: evidence of neutron emission and Non-radioactiv product elements. Exp Mech 53:345–365
28. Aggelis DG, Soulioti DV, Sapouridis N, Barkoula NM, Paipetis AS, Matikas TE (2011 Acoustic emission characterization of the fracture process in fibre reinforced concrete. Con struct Build Mater 25:4126–4131
29. Aggelis DG, Mpalaskas AC, Matikas TE (2013) Acoustic signature of different fracture mode in marble and cementitious materials under flexural load. Mech Res Commun 47:39–43
30. Scholz CH (1968) The frequency-magnitude relation of microfracturing in rock and its relatio to earthquakes. Bull Seismol Soc Am 58:399–415
31. Carpinteri A, Lacidogna G, Pugno N (2006) Richter's laws at the laboratory scale interprete by acoustic emission. Mag Concr Res 58:619–625
32. Niccolini G, Carpinteri A, Lacidogna G, Manuello A (2011) Acoustic emission monitoring o the Syracuse Athena temple: scale invariance in the timing of ruptures. Phys Rev Let 108503:1–4
33. Carpinteri A, Lacidogna G, Borla O, Manuello A, Niccolini G (2012) Electromagnetic an neutron emissions from brittle rocksfailure: experimental evidence and geological implica tions. Sadhana 37:59–78
34. Volodichev NN, Kuzhevskij BM, Nechaev O, Yu PMI, Podorolsky AN, Shavrin PI (2000 Sun-moon-earth connections: the neutron intensity splashes and seismic activity, astron. Vest 34:188–190

Chapter 3
Neutron Emissions and Compositional Changes at the Compression Failure of Iron-Rich Natural Rocks

Amedeo Manuello, Riccardo Sandrone, Salvatore Guastella, Oscar Borla, Giuseppe Lacidogna, and Alberto Carpinteri

Abstract Neutron emissions (NE) were measured during laboratory experiments conducted on iron-bearing and iron-rich rocks. In particular, magnetite specimens were loaded up to the final failure under monotonic displacement control. Also basalt rocks were tested under cyclic loading conditions (2 Hz) up to the final failure. In order to detect neutron emissions, the tests were monitored by two different neutron measurement devices: He^3 proportional counter and thermodynamic (bubble) detectors. After the experiments, Energy Dispersive X-Ray Spectroscopy (EDS) analyses were carried out to detect possible direct evidences of low energy nuclear reactions (piezonuclear fission reactions) on the fracture surfaces. In particular, quantitative evidences of nuclear reactions, involving iron decrease and the corresponding increase in lighter elements, were observed in the olivine, crystalline mineral phase widely diffused in the basalt matrix, and in the magnetite. These results reinforce the evidences previously observed for Luserna stone (granitic orthogneiss) and confirm that piezonuclear fission reactions take place in natural iron-bearing materials subjected to damage accumulation and cracking.

Keywords Neutron emission • Rocks fracture • X-Ray spectroscopy • Geochemical evolution

A. Manuello (✉) • O. Borla • G. Lacidogna • A. Carpinteri
Department of Structural, Geotechnical and Building Engineering, Politecnico di Torino, Corso Duca degli Abruzzi 24, 10129 Torino, Italy
e-mail: amedeo.manuellobertetto@polito.it

R. Sandrone
Department of Environment, Land and Infrastructure Engineering, Politecnico di Torino, Corso Duca degli Abruzzi 24, 10129 Torino, Italy

S. Guastella
Department of Applied Science and Technology, Politecnico di Torino, Corso Duca degli Abruzzi 24, 10129 Torino, Italy

© Springer International Publishing Switzerland 2015
A. Carpinteri et al. (eds.), *Acoustic, Electromagnetic, Neutron Emissions from Fracture and Earthquakes*, DOI 10.1007/978-3-319-16955-2_3

3.1 Introduction

It is possible to demonstrate experimentally that brittle fracture in solid materials can be accompanied by the release of different forms of energy [1–3]. In recent studies, it has been observed that quasi-brittle materials such as granitic orthogneiss (Luserna stone) subjected to compression tests under monotonic displacement control, by cyclic loading, or by ultrasonic vibration, are characterized by neutron emissions up to one order of magnitude greater than the background level [1–11]. These tests were conducted on Luserna stone specimens with different shapes and dimensions and characterized by an iron oxide content of approximately 1.5 %. In the case of this rock, the iron oxides are prevalently concentrated within two minerals: phengite and biotite. These two minerals, rather common in such a stone (up to 20 % and 2 %, respectively in volume), have shown important changes in the mineral chemistry of the fracture surfaces after the experiments [8–11]. The reduction in Fe content (~25 %) seems to be almost perfectly compensated by an increment in Al, Si, and Mg [8–11].

In the present investigation, neutron emission measurements, by means of a He^3 proportional counter and thermodynamic bubble detectors, were performed during compression tests on magnetite specimens and during cyclic loading tests carried out on basalt rocks. The employed materials have been chosen in order to correlate the iron contents (~15 % for basalt and ~72.5 % for magnetite) to the neutron emission levels measured during crushing and fatigue tests.

The crushing tests on magnetite were performed using cylindrical specimens of different diameters coming from the San Leone mine. This deposit is located at about 30 km southwest of Cagliari, Sardinia (Italy), near the Basso-Sulcis batholit, a granodioritic intrusive of the Hercynian age [12]. This mine is rather recent, it was discovered in 1860 during the industrial development [12]. In 1892, the mine started its activity, which ended in 1963. The mineralization of the deposit is mainly represented by magnetite. The ore bodies within skarn were generated by the thermometamorphism of previous paleozoic limestones in contact with granitic intrusions of Variscan age (300×10^6 years ago). The mean iron concentrations in the San Leone magnetite is between 72.5 % and 75 % [12].

Another iron-rich material used for the cyclic loading experiments is a basalt coming from the Mount Etna. Basalt, a very common extrusive igneous rock, is the dominant material making up the oceanic Earth's Crust and represents the principal product of volcanic eruptions. Basalt is characterized by a mafic chemistry (with a high content of iron and magnesium), is dark grey and shows porphyritic texture with few mm-sized phenochrists of plagioclase, clinopyroxene and olivine [13].

After the experiments, the basalt and magnetite fracture surfaces were analyzed by Energy Dispersive X-ray Spectroscopy (EDS) in order to obtain a direct evidence of piezonuclear fission reactions that could have taken place during the tests. The results confirm the evidence observed in the case of Luserna stone and permit to recognize different piezonuclear reactions induced in natural non-radioactive rocks characterized by a certain iron concentrations.

As far as the magnetite is concerned, the EDS analysis showed an impressive direct evidence of piezonuclear fission reactions. After the magnetite experiments, macroscopical changes have been observed, comparing external and fracture surfaces. Among these changes, the appearance of appreciable quantities of Al, previously absent, seems to be particularly significant. The chemical changes in basalt and magnetite could be considered to give a valid interpretation to the different chemical compositions between the oceanic and the continental crust. In other terms, we could explain the transition from basaltic to sialic chemical composition that characterized the geochemical evolution of the Earth's Crust [14, 15].

3.2   Granitic Orthogneiss (Luserna Stone): Early Experiments

Preliminary tests on prismatic specimens were presented in previous contributions, recently published [1–3], and related to piezonuclear reactions occurring in solids containing iron (samples of Luserna stone in compression). In a preliminary experiment, four specimens were tested, two made of Carrara marble and two made of Luserna stone (Fig. 3.1). All of them were of the same size and shape, measuring $6 \times 6 \times 10$ cm^3. The neutron measurements obtained on the two Luserna stone specimens exceed the background level by approximately one order of magnitude, when catastrophic failure occurred. The first specimen reached at time T = 32 min a peak load of ca 400 kN, corresponding to an average pressure on the bases of 111.1 MPa. When failure occurred, the count rate was found to be $(28.3 \pm 5.7) \times 10^{-2}$ cps, corresponding to an equivalent flux of thermal neutrons of $(43.6 \pm 8.8) \times 10^{-4}$ $n_{thermal}$ cm^{-2} s^{-1} (see Fig. 3.2). The second specimen reached at time T = 29 min a peak load of ca 340 kN, corresponding to an average pressure on the bases of 94.4 MPa. When failure occurred, the count rate was found to be $(27.2 \pm 5.5) \times 10^{-2}$ cps, corresponding to an equivalent flux of thermal neutrons of $(41.9 \pm 8.5) \times 10^{-4}$ $n_{thermal}$ cm^{-2} s^{-1}.

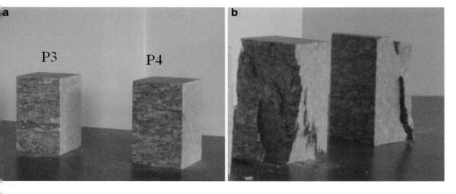

Fig. 3.1 Granitic orthogneiss (Luserna stone) specimens before (**a**) and after the fracture tests (**b**)

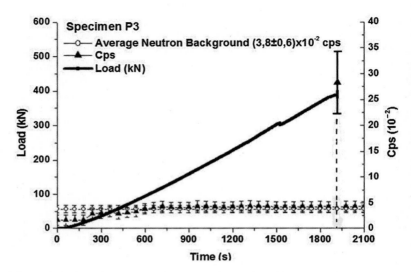

Fig. 3.2 Load vs. time and cps curve for test specimen of granitic orthogneiss

In more recent experiments, cylindrical specimens with different size and slenderness were selected, instead of prismatic specimens as in the preliminary tests. For the specimens of larger dimensions, neutron emissions, detected by He^3 were found to be of about one order of magnitude higher than the ordinary natural background level at the time of the catastrophic failure. These emissions fully confirmed the preliminary tests [1–10]. For specimens with sufficiently large size and/or slenderness, a relatively high energy release is expected, and hence a higher probability of neutron emissions at the time of failure. Furthermore, during compression tests with cyclic loading, an equivalent neutron dose was found at the end of the test by neutron bubble detectors, about twice higher than the ordinary background level [4, 8, 9]. In addition, using an ultrasonic horn suitably joined with the specimen, ultrasonic tests were carried out on Luserna stone producing a continuous vibration at 20 kHz. At the end of the test, an equivalent neutron dose about twice higher than the background level was detected [4, 5]. In the present paper, in order to evaluate the correlation between iron content and neutron emission, specific tests are described on basalt and magnetite.

3.3 Basalt and Magnetite Specimens: Experimental Set-Up and Neutron Emission Evidences

Experimental compression tests were performed on brittle magnetite specimens under monotonic displacement control, and on basaltic specimens under cyclic loading. All the experimental tests were realized at the Fracture Mechanics

Table 3.1 Tested materials and test types

Monotonic displacement control						
Specimen	Material	D (mm)	H (mm)	$\lambda = H/D$	Velocity (mm/s)	Load (kN)
P1	Magnetite	40	40	1	0.001	197.97
P2	Magnetite	90	90	1	0.002	932.38
P3	Magnetite	20	20	1	0.0005	45.05
Cyclic loading						
Specimen	Material	D (mm)	H (mm)	$\lambda = H/D$	Frequency (Hz)	Maximum load (kN)	Minimum load (kN)
P4	Basalt	80	160	2	2	350	30

Laboratory of the Politecnico di Torino. In particular, in this paper the experimental results of four specimens P1, P2, P3 (magnetite) and P4 (basalt) are reported. The tested materials, the shapes and sizes of the specimens, the characteristics of the testing procedure are summarized in Table 3.1.

The crushing experiments on magnetite were performed in uniaxial compression, utilizing a MTS servo-controlled hydraulic testing machine with a maximum capacity of 1000 kN. Each test was performed in piston travel displacement control by setting a constant piston velocity between 5×10^{-4} and 1×10^{-3} mm/s. The test specimens were arranged in contact with the press platens without any coupling materials, according to the testing modalities known as "test by means of rigid platens with friction". The fatigue test on the basalt specimen was performed up to the final failure at a frequency of 2 Hz, with a maximum load of 350 kN and a minimum load of 30 kN (Table 3.1).

The neutron emission measurements on magnetite specimens were performed using an He^3 neutron detector switched on at least 1 h before the beginning of each compression test, in order to reach the thermal equilibrium of electronics, and to make sure that the behaviour of the device was stable with respect to intrinsic thermal effects. The average measured background level was ranging from $(4.00 \pm 1.00) \times 10^{-2}$ to $(6.40 \pm 1.60) \times 10^{-2}$ cps. Neutron measurements of specimen P3 yielded values comparable with the ordinary natural background, whereas in specimen P1 the experimental data exceeded the background level by approximately five times. For specimen P2 the neutron emissions achieved values of about three orders of magnitude higher than the ordinary background (Fig. 3.3a). Specimen P4, made of basaltic rock, was subjected to fatigue cycles up to the final failure as summarized in Table 3.1. Droplets counting was performed every 12 h and the equivalent neutron dose was calculated [8, 9, 16]. In the same way, the natural background was estimated by means of the two bubble dosimeters. During this test the ordinary background was found to be (53.76 ± 13.44) nSv/h. An increment of about 20 times with respect to the background level was detected at specimen failure (Fig. 3.3b). No significant variations in neutron emissions were observed before the failure. The equivalent neutron dose, at the end of the test, was (935.49 ± 233.87) nSv/h.

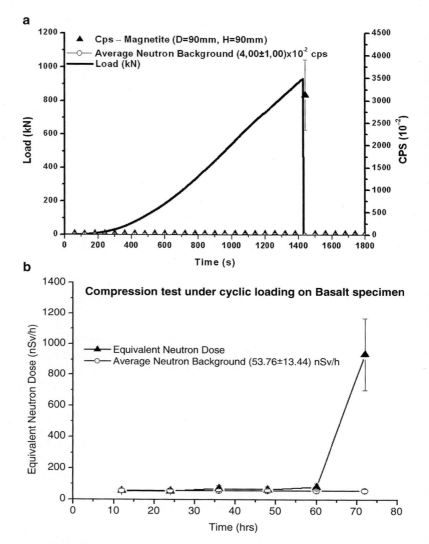

Fig. 3.3 (**a**) Magnetite: For specimen P2 the neutron emissions achieved values of about three orders of magnitude higher than the background level. (**b**) Basalt: An increment of about 20 times in comparison with the background level was detected at the failure of specimen P4 (fatigue test)

3.4 Compositional Changes: EDS Analysis on Basalt

After the mechanical loading experiments, Energy Dispersive X-ray Spectroscopy (EDS) was performed on different samples of external and fracture surfaces, belonging to the same specimens used during the cyclic loading test on basalt.

Fig. 3.4 (**a**) Polished thin sections, finished with a standard petrographic sample procedure, covered by Cr, were examined to evaluate olivine composition on external surface; (**b, c**) small portions of fracture surfaces without any kind of preparation, apart from the Cr covering, were analysed to evaluate the chemical changes in olivine on fracture surface

The measurement precision is in the order of magnitude of 0.1%. The analysis was conducted in order to correlate the neutron emission from the specimen with the variations in rock composition and to detect possible anomalous transformations from iron to lighter elements. The quantitative elemental analyses were performed by a ZEISS Supra 40 Field Emission Scanning Electron Microscope (FESEM) equipped with an Oxford X-rays microanalysis [8–11].

The first analysis was performed on fracture surfaces of basalt specimen P4 after the fatigue test and the consequent failure. In this case, similarly to the case of Luserna stone [11], taking into account the heterogeneity of the material, the samples were carefully chosen to investigate and compare the same minerals before and after the failure. In particular, olivine was considered due to its high iron content (~24 %) and because it is rather abundant within this type of rock [11]. In the case of basalt, two different kinds of samples were examined: (i) polished thin sections, finished with a standard petrographic sample procedure, covered by Cr, for what concerns the external surface of the basalt; (ii) small portions of fracture surface without any kind of preparation, apart from the Cr covering, for what concerns the fracture surface of the same material. A semi-quantitative nonstandard analysis was performed on the collected spectra, fixing the stoichiometry of the oxides, to correlate the oxides content with the specific crystalline phase. The Cr lines were excluded from the semi-quantitative evaluation [11].

In Figs. 3.4a–c, a polished thin section obtained from the external surface of an integer and un-cracked portion of the basalt specimen P4 are shown together with two fragments of fracture surface. The polished thin section presents a rectangular geometry (45×27) mm^2 and is 30 μm thick. This kind of analysis involves millions of cubic microns for each acquisition area. From this point of view, there exists a substantial difference between these analyses and the spot analysis reported for the Luserna stone samples [11]. In this case, in fact, a larger portion of material is involved and each analysis is indicative of an investigated volume equal to $60 \times 20 \times 4\ \mu m^3$.

In Figs. 3.5a–c, the distributions of Fe, Si and Mg concentrations in olivine are reported for external and fracture surfaces of specimen P4. It can be observed that the distribution of Fe content for the external surface shows an average value of 18.4 % (Fig 3.5a). In the same graph, the distribution of Fe concentrations on the fracture sample shows a significant variation. It can be seen that the mean value of the distribution of measurements performed on the fracture surface is equal to 14.4 %, considerably lower than the mean value of external surface measurements (18.4 %). Similarly to Fig. 3.5a, in Fig 3.5b the Si mass percentage concentrations are considered. For Si contents, the observed variations show a mass percentage increase approximately equal to 2.2 %. The average value of Si concentrations changes from 18.3 % on the external surface to 20.5 % on the fracture surface. In Fig. 3.5c it is shown that, in the case of olivine, also Mg content presents considerable variations. Fig. 3.5c shows that the mass percentage concentration of Mg changes from a mean value of 21.2 % (external surface) to a mean value of 22.8 % (fracture surface) with an increase of 1.6 %. Therefore, the iron decrease (−4.0 %) in olivine seems to be almost perfectly counterbalanced by an increase in silicon (+2.2 %), and magnesium (+1.6 %). A further analysis conducted on olivine and localized on the fracture surface shows the appearance of Al_2O_3 in the chemical composition of this crystalline phase (see Table 3.2). This evidence seems to be particularly important because no Al traces were observed in the olivine sample localized on the external surface. This fact represents a further confirmation that in basaltic olivine, similarly to the case of phengite in Luserna stone [11], the following piezonuclear reaction occurred:

$$Fe_{26}^{56} \rightarrow 2Al_{13}^{27} + 2\ \text{neutrons} \qquad (3.1)$$

At the same time, the results involving the Fe decrease and the consequent increases in Si and Mg, discussed above and reported in Fig. 3.5, lead to the conclusion that in olvine, as well as in the biotite of Luserna stone [11], the following piezonuclear reaction occurred during the mechanical loading:

$$Fe_{26}^{56} \rightarrow Mg_{12}^{24} + Si_{14}^{28} + 4\ \text{neutrons} \qquad (3.2)$$

The results reported in Figs. 3.5 represent also an important significance from a geophysical point of view. In fact, at the scale of the Earth's Crust, the non-homogeneous composition of oceanic and continental crusts could be explained by the transition from basaltic to sialic compositions. Comparing the data presented in the literature concerning the composition of the two different types of terrestrial crust it can be noted that the iron concentration changes from ~8 %, in the oceanic crust, to ~4 % in the continental one [17–22]. Ni changes from ~0.03 %, in the oceanic crust, to ~0.01 % in the continental one (about a three-fold decrease). And vice versa, Al, Si

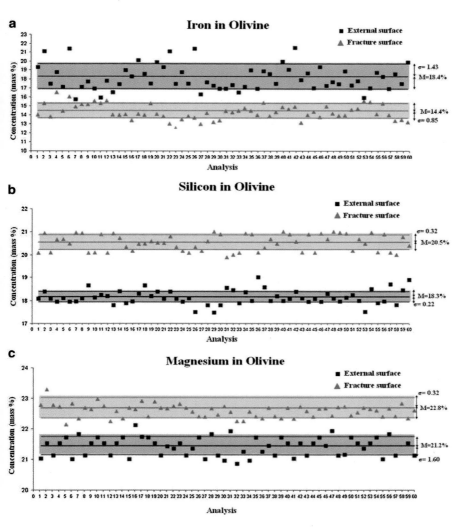

Fig. 3.5 Olivine chemical changes after mechanical loading: (**a**) The Fe decrease (−4.0 %) is almost perfectly counterbalanced by an increase in Si (**b**) (+2.2 %), and Mg (**c**) (+1.6 %).

Mg, and Na vary from ~7 %, ~24 %, ~3.6 %, and ~1 % in the oceanic crust, to ~8 %, ~28 %, ~1.3 %, and ~2.9 % in the continental crust, respectively. Considering that approximately 50 % of the continental crust has originated over the last 3.8 Gyrs, as the result of oceanic crust subduction [3–5, 14, 15, 18–22], the results presented in this paper offer a further confirmation that piezonuclear reactions are a possible explanation for the chemical changes in the crust in correspondence of mechanical phenomena of fracture, crushing, fragmentation, comminution, erosion, friction, due to seismic and tectonic events [1–5, 9, 14, 15].

Table 3.2 Olivine: Fe, Si, Mg, and Ca weight percentage mean values on external and fracture surfaces. Variations with respect to the mineral and to the same element

	External surface mean value (wt%)	Fracture surface mean value (wt%)	Increase/decrease with respect to the mineral	Increase/decrease with respect to the element itself
Fe	18.4	14.4	−4.0 %	−21 %
Si	18.3	20.5	+2.2 %	+12 %
Mg	21.2	22.8	+1.6 %	+7 %
Ca	0.5	0.5	No variations	No variations

3.5   EDS Analysis on Magnetite

Similar quantitative results have been obtained also in the case of magnetite experiments. The fracture surfaces of specimen P2 were analyzed in order to recognize possible evidence of piezonuclear reactions. Taking into account the homogeneity of the chemical composition of this material (Fe-oxides content ~95 %), the analysis on the external surface of this rock was conducted in a different manner with respect to granitic orthogneiss and basalt. In this case, in fact, taking into account the homogeneity of the rock, both map and spot analyses were used to evaluate the changes in element distribution and, eventually, the appearance of other elements, previously absent. As mentioned before, the typical composition of magnetite is given by Fe and O, these elements representing about the 95 % of this kind of rock. The remaining part is represented by traces of other elements such as Na, Si, Cl, and K. Maps of dimension 250 μm × 200 μm were analyzed to localize the presence of significant changes in the chemical composition.

In Figs. 3.6 and 3.7, the Fe and Al concentrations together with the concentrations of the other elements are reported indicating the mass percentage concentrations in the case of both external and fracture surface. At a first glance, it can be noted that, for the fracture surface, a consistent Al content appears after the specimen failure (Fig. 3.7e). The first important evidence is given by the fact that the presence of Al and Mn, absent in the analysis performed on the external surface, were observed in several points localized on the analyzed fracture surface (see Figs. 3.7e, 3.7f).

At the same time, on the fracture surface, we observe a significant Fe decrease together with the increase in elements such as Si and O. In order to evaluate these changes from a statistical point of view, 15 analyses on the external surface and 15 on the fracture surface were carried out and reported in Fig. 3.8 and in Figs. 3.9a–e. In these diagrams, similarly to the results obtained for olivine and reported in the previous section, the mass percentage concentrations for Fe (Fig. 3.8), Al, Mn, Si, and O (Fig. 3.9) are reported making a comparison between external and fracture surfaces. It is interesting to observe a decrease in Fe concentration of 27.9 % starting from a value of 64.8 % (external surface) and going down to a concentration of 36.9 % after the experiment (fracture surface) (Fig. 3.8). At the same time,

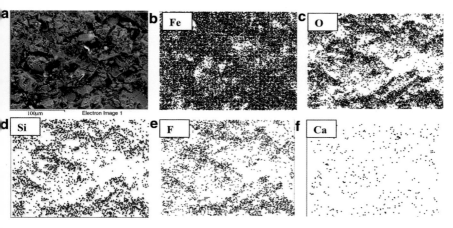

Fig. 3.6 Maps obtained from EDS analysis of magnetite (**a**). The maps, with dimensions of about 250 μm × 200 μm, were analyzed before the loading tests. The typical concentration on external surfaces show high percentages of Fe (~65 %) (**b**) and O (~30 %) (**c**). The remaining concentration s represented by minor contents of Si (**d**), F (**e**) and Ca (**f**)

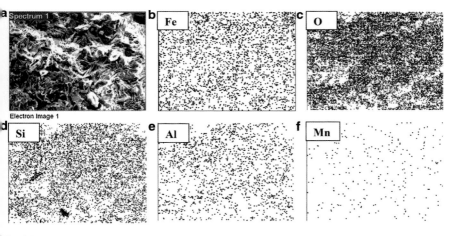

Fig. 3.7 Maps obtained from EDS analysis of fracture surface (**a**). The maps, with dimensions of about 250 μm × 200 μm, were analyzed to localize the presence of significant changes in the chemical composition. The typical concentration on fracture surfaces shows a lower percentage of Fe (~35 %) (**b**). The concentration of O reached a content of ~38 % (**c**). At the same time an appreciable Si concentration (~10 %) has been observed after the tests (**d**). The impressive evidence regarded the appearance of Al (**e**) and Mn (**f**) after the experiments with a mass percentage of about 10 % and 2.2 % respectively

Al concentration increases from zero to a mean value of 10.1 %. It is interesting to consider that approximately one third of the Fe decrease may be counterbalanced by the Al increase, according to reaction (1). The remaining part of the Fe decrease seems to be almost perfectly counterbalanced by the increase in the other elements.

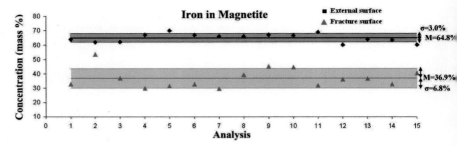

Fig. 3.8 Fe decreased by about 27.9 % in magnetite after the brittle fracture of the specimen. The Fe concentration on the external surface was about 64.8 % and the concentration on the fracture surface, after the test, was 36.9 %

In fact, the appearance of Mn (+2.2 %), the increase in Si (+8.7 %), and the increase in O (+7.6 %) may be considered in the light of the following reactions (see Table 3.3):

$$\mathrm{Fe}_{26}^{56} \rightarrow \mathrm{Mn}_{25}^{55} + \mathrm{H}_1^1 \tag{3.3}$$

$$\mathrm{Fe}_{26}^{56} \rightarrow \mathrm{Si}_{14}^{28} + \mathrm{O}_8^{16} + 2\mathrm{He}_2^4 + 4 \ \text{neutrons} \tag{3.4}$$

The appearance of Al and Mn, previously absent, and the almost perfect balance between the Fe decrease (-27.9 %) and the increase in the other elements (+27.7 %), well-matched by piezonuclear fissions (3.1), (3.3) and (3.4), could be considered as direct evidences that nuclear reactions of a new type take place.

3.6 Conclusions

We report the results of neutron bursts measured during the application of monotonic and cyclic loading to Sardinian magnetite and Etna basalt specimens. These analyses are strictly related to recent results obtained from similar tests on inert rocks with a lower Fe content (Luserna stone).

In the cases reported in this paper, the piezonuclear fission reactions regarding Fe as the starting element are even more evident. These investigations confirm that high-frequency pressure waves, suitably exerted on inert and stable nuclides, generate nuclear reactions of a new type (piezonuclear fission reactions), with substantial chemical changes in the mineral compositions of iron-rich rocks. EDS analyses were performed on different samples of external or fracture surface belonging to basalt specimens, in order to get an average information about possible changes in the chemical composition of olivine. Considering the results for this crystalline phase, a significant decrement in the iron content (-4 %) is counterbalanced by the increment in lighter elements such as Si and Mg.

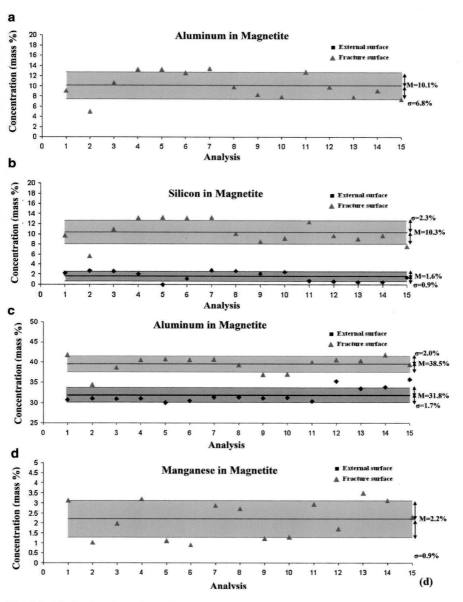

Fig. 3.9 Al, Si, O and Mn increasing on magnetite fracture surface after the experiments. In particular Al and Mn, with concentrations of about 10.1 % and 2.2 % after the experiments, were absent before the test or observable in trace on the external surfaces

The results reported for magnetite are even more evident, with the appearance of elements lighter than iron, previously absent. In this case, the decrease in Fe concentration is nearly equal to 30 %, and the appearance of Al and Mn, previously

Table 3.3 Magnetite: Fe, Al, Mn, Si and O weight percentage mean values on external and fracture surfaces. Variations with respect to the mineral material and to the same element

	External surface mean value (wt %)	Fracture surface mean value (wt %)	Increase/decrease with respect to the mineral	Increase/decrease with respect to the element itself
Fe	64.8	36.8	−28.0 %	−56 %
Al	−	10.1	+10.1 %	Before absent
Mn	−	2.2	+2.2 %	Before absent
Si	1.6	10.3	+8.7 %	+540 %
O	31.8	38.5	+6.7 %	+21 %

absent, represents a very impressive result. In particular for Al, the concentration on fracture surfaces reached, starting from zero, a mean value of about 10 %. The remaining part of the Fe decrement seems to be almost perfectly counterbalanced by the increments in Si and O.

The results coming from EDS analysis on magnetite are particularly significant, considering the theoretical explanations to these phenomena recently given by Widom et al. [23, 24], in order to explain neutron emissions as a consequence of nuclear reactions taking place in iron-rich rocks during brittle fracture. The same authors argued that neutron emissions may be related to piezoelectric effects and that the fission of iron may be a consequence of the nuclear photodisintegration [23].

As a conclusion and according to the experimental results, the hypothesis of piezonuclear reactions regarding iron depletion finds a very important evidence and confirmation at the Earth's Crust scale. The piezonuclear reactions have thus been considered in order to interpret the most significant geophysical and geological transformations, today still unexplained. The laboratory results herein reported for basalt and magnetite give a new and original interpretation to the evolution from the Hadean to the Archean lithosphere of our planet, and contribute to explain the difference between oceanic and continental crust compositions due to subductive and seismic phenomena.

Acknowledgements Special thanks are due to Prof. R. Ciccu and Mr. F. Argiolas for providing the magnetite specimens.

References

1. Carpinteri A, Cardone F, Lacidogna G (2009) Piezonuclear neutrons from brittle fracture: early results of mechanical compression tests. Strain 45:332–339, Atti dell'Accademia delle Scienze di Torino, Torino, Italy, 33: 27–42
2. Cardone F, Carpinteri A, Lacidogna G (2009) Piezonuclear neutrons from fracturing of inert solids. Phys Lett A 373:4158–4163

3. Carpinteri A, Cardone F, Lacidogna G (2010) Energy emissions from failure phenomena: mechanical, electromagnetic, nuclear. Exp Mech 50:1235–1243
4. Carpinteri A, Borla O, Lacidogna G, Manuello A (2010) Neutron emissions in brittle rocks during compression tests: monotonic vs. cyclic loading. Phys Mesomech 13:268–274
5. Carpinteri A, Lacidogna G, Manuello A, Borla O (2011) Energy emissions from brittle fracture: neutron measurements and geological evidences of piezonuclear reactions. Strength, Fracture Complex 7:13–31
6. Carpinteri A, Lacidogna G, Manuello A, Borla O (2011) Piezonuclear transmutations in brittle rocks under mechanical loading: microchemical analysis and geological confirmations. In: Kounadis AN, Gdoutos EE (eds) Recent advances in mechanics. Springer, Chennai, pp 361–382
7. Carpinteri A, Lacidogna G, Borla O, Manuello A, Niccolini G (2012) Electromagnetic and neutron emissions from brittle rocks failure: experimental evidence and geological implications. Sadhana 37:59–78
8. Carpinteri A, Lacidogna G, Manuello A, Borla O (2012) Piezonuclear fission reactions: evidences from microchemical analysis, neutron emission, and geological transformation. Rock Mech Rock Eng 45:445–459
9. Carpinteri A, Lacidogna G, Manuello A, Borla O (2013) Piezonuclear fission reactions from earthquakes and brittle rocks failure: evidence of neutron emission and nonradioactive product elements. Exp Mech 53(3):345–365
10. Carpinteri A, Borla O, Lacidogna G, Manuello A (2012) Piezonuclear reactions produced by brittle fracture: from laboratory to planetary scale. In: Proceedings of the 19th European conference of fracture, Kazan, Russia
11. Carpinteri A, Chiodoni A, Manuello A, Sandrone R (2011) Compositional and microchemical evidence of piezonuclear fission reactions in rock specimens subjected to compression tests. Strain 47(2):267–281
12. Verkaeren J, Bartholomè P (1979) Petrology of the San Leone magnetite skarn deposit (S. W. Sardinia). Econ Geol 74:53–66
13. Tanguy JC (1978) Tholeiitic basalt magmatism of Mount Etna and its relations with the alkaline series. Contrib Mineral Petrol 66:51–67
14. Carpinteri A, Manuello A (2011) Geomechanical and geochemical evidence of piezonuclear fission reactions in the Earth's crust. Strain 47:282–292
15. Carpinteri A, Manuello A (2012) An indirect evidence of piezonuclear fission reactions: geomechanical and geochemical evolution in the Earth's crust. Phys Mesomech 15:14–23
16. Bubble Technology Industries (1992) Instruction manual for the bubble detector. Bubble Technology Industries, Chalk River, ON
17. Catling CD, Zahnle KJ (2009) The planetary air leak. Sci Am 300:24–31
18. Taylor SR, McLennan SM (1995) The geochemical evolution of the continental crust. Rev Geophys 33:241–265
19. Taylor SR, McLennan SM (2009) Planetary crusts: their composition origin and evolution. Cambridge University Press, Cambridge
20. Fowler CMR (2005) The solid earth: an introduction to global geophysics. Cambridge University Press, Cambridge
21. Doglioni C (2007) Interno della Terra, Treccani, Enciclopedia. Scienza e Tecnica, 595605
22. Rudnick RL, Fountain DM (1995) Nature and composition of the continental crust: a lower crustal perspective. Rev Geophys 33:267–309
23. Widom A, Swain J, Srivastava YN (2015) Photo-disintegration of the iron nucleus in fractured magnetite rocks with magnetostriction. Meccanica 50:1205–1216
24. Widom A, Swain J, Srivastava YN (2013) Neutron production from the fracture of piezoelectric rocks. J Phys G Nucl Part Phys 40:15006 (1–8)

Chapter 4
Frequency-Dependent Neutron Emissions During Fatigue Tests on Iron-Rich Natural Rocks

Alberto Carpinteri, Francesca Maria Curà, Raffaella Sesana, Amedeo Manuello, Oscar Borla, and Giuseppe Lacidogna

Abstract The results coming from neutron emission measurements during fatigue experiments performed at low (2 Hz), intermediate (200 Hz), and high (20 kHz) frequency are reported. These results confirm that appreciable neutron emissions, greater than the background level, may be observed during damage accumulation in iron-bearing rocks: granite (Fe oxides ~1.5 %), basalt (Fe oxides ~15 %), and magnetite (Fe oxides ~75 %). The neutron detection, together with temperature measurements obtained by infrared revelation, lead to the conclusion that fatigue tests performed at 200 Hz represent the condition for which the neutron emission is the highest. This evidence seems to be particularly important considering recent results from seismological observation of the Jacinto fault in the Southern California and the neutron emissions detected during seismic activity.

Keywords Fatigue experiments • Neutron emissions • Luserna stone • Basalt • Magnetite

4.1 Introduction

In the last few years, neutron emissions have been observed in several experiments characterized by static or repeated (fatigue) loading test condition on inert iron-rich rocks by Carpinteri et al. [1–8]. The large amount of these evidences suggests that pressure, suitably exerted on inert and stable nuclides, generates nuclear reactions of a new type, producing energy emission in the form of neutrons [1–3]. In particular, in several cases, neutron emission measurements, by means of He^3

A. Carpinteri (✉) • A. Manuello • O. Borla • G. Lacidogna
Department of Structural, Geotechnical and Building Engineering,
Politecnico di Torino, Corso Duca degli Abruzzi 24, 10129 Torino, Italy
e-mail: alberto.carpinteri@polito.it

F.M. Curà • R. Sesana
Department of Mechanical And Aerospace Engineering, Politecnico di Torino,
Corso Duca degli Abruzzi 24, 10129 Torino, Italy

© Springer International Publishing Switzerland 2015
A. Carpinteri et al. (eds.), *Acoustic, Electromagnetic, Neutron Emissions from Fracture and Earthquakes*, DOI 10.1007/978-3-319-16955-2_4

devices and thermodynamic bubble detectors, were performed during different kinds of experiments: monotonic compressive and tensile tests, cyclic loading, ultrasonic vibrations.

Very recently, a theoretical interpretation was proposed by Widom et al. explaining neutron emissions as a consequence of nuclear fission reactions taking place in iron-rich rocks during brittle micro-cracking or fracture [9, 10]. It was observed that iron nuclear disintegration takes place when rocks containing such nuclei are crushed and fractured. The resulting nuclear trasmutations are particularly relevant in the case of magnetite and iron-rich materials in general. The same authors argued that neutron emissions may be related to piezoelectric effects and that the fission of iron may be a consequence of photodisintegration of the nuclei [9].

In the experiments reported in previous publications, granitic ortho-gneiss specimens showed a maximum neutron emission approximately ten times higher than the measured background level. The emission was $(28.3 \pm 5.7) \times 10^{-2}$ cps, against a measured background noise of about $(3.8 \pm 0.6) \times 10^{-2}$ cps [3–8]. At the same time, Etna basalt, used during similar tests, produced neutron bursts with an equivalent dose rate up to 10^2 times the average background level in correspondence to the catastrophic compression failure.

The most impressive result is that of magnetite specimens, characterized by a very high content of iron (~75 %). They were subjected to crushing experiments and reached a neutron emission level (935.49 ± 233.87) nSv/h up to 10^3 times the background level (5.76 ± 13.44) nSv/h. The results obtained in the last few years from fatigue tests performed at low and high frequency are herein extended to an intermediate frequency, around 200 Hz, by means of a electromechanical Amsler Vibrofore. This equipment allows to apply fatigue loading with a frequency range between 100 and 260 Hz, a mean load up to 50 kN, and an alternate load up to 50 kN. As reported in two previous papers by Carpinteri et al., the cyclic tests, using 2 Hz or 20 KHz as working frequency, were obtained by a servo-hydraulic press for low frequency and by a Bandelin HD 2200 sonotrode for ultrasounds [4, 5]. At a first glance, from the experiments reported in the literature, it was possible to correlate neutron emissions and consequent nuclear transmutations (compositional changes) [3–8]. By working on the results coming from the experiments conducted on natural rocks in static and cyclic-fatigue loading, it has been observed that the greater the iron content, the higher the neutron emissions [1–8]. From the same experiments, the neutron level may be also correlated to the size-scale of the specimen. As a matter of fact, the greater the size-scale and the slenderness, the higher the neutron bursts.

In the present paper, Luserna stone (a granitic rock), Basalt, and Magnetite, containing different iron concentrations, were tested in fatigue experiments using low (2 Hz), intermediate (200 Hz), or high (20 kHz) frequency. We investigated different cyclic conditions to confirm that energy emission in the form of neutrons takes place also in the case of fatigue tests under different frequency regimes. The experiments had also the purpose to give a first indication about a possible correlation between neutron emission and loading frequency.

4.2 Experimental Set Up

4.2.1 Materials

The specimens were obtained from different iron-bearing rocks. The tests were conducted on Luserna stone (granitic ortho-gneiss) characterized by a content in iron-oxide of about 1.5 %, on basalt with an iron content of ~15 %, and on magnetite with an iron content of ~90 %, respectively. The employed materials were chosen in order to confirm the correlation between this element concentration and the neutron emission levels. During the experimental campaign, as stated in the introduction, the specimens were subjected to fatigue cycles with different working frequencies. More precisely, the test frequencies were 2 Hz, 200 Hz, and 20 kHz. The main purpose was that of detecting neutron emission activity during the tests in order to correlate it with the frequency range, as well as with the iron content and the specimen size.

From a petrographic point of view, the Luserna stone used for the experimental campaign is a leucogranitic orthogneiss, probably from the Lower Permian Age, that outcrops in the Luserna-Infernotto basin (Cottian Alps, Piedmont), at the border between the Turin and Cuneo provinces (north-western Italy) [11–14]. The rock is characterised by a micro'Augen' texture, it is grey-greenish or locally pale blue in colour. Geologically, Luserna stone pertains to the Dora-Maira Massif [12, 13], that represents a part of the ancient European continent during Alpine orogenesis [11, 13]. The second material employed during the experimental study is basalt originating from the Mount Etna. Mount Etna is composed for the most part of intermediate alkaline products, that may be defined as sodic trachybasalts or trachyandesites. The strato-volcanio itself overlies tholeiitic basalts [15].

Basalt is a very common extrusive igneous rock, the dominant material making up the Earth's oceanic crust, and represents the principal product from volcanic eruptions. The basalt used in the tests reported in this paper, is characterized by a mafic composition (with a high content of iron and magnesium), is dark grey and shows porphyritic texture with few mm-sized phenochrists of plagioclase, clinopyroxene and olivine [15]. Magnetite specimens come from the San Leone mine. This deposit is located about 30 km southwest of Cagliari, Sardinia (Italy), near the Basso-Sulcis batholit, a granodioritic intrusive of Hercynian age [16]. The mineralization of the deposit is mainly represented by magnetite. The ore bodies within skarn generated by the thermometamorphism of previous paleozoic limestones in contact with granitic intrusions of Variscan age. The mean iron concentrations of the San Leone magnetite is between 72.5 % and 75.0 % [16]. All the prepared specimens were cylindrical, with a diameter of 50 mm and a lenght of 100 mm. Figures 4.1a–c show three different specimens representing the three types of tested rock. The geometrical and mechanical characteristics of Luserna stone, basalt, and magnetite specimens are summarized in Table 4.1.

Fig. 4.1 Luserna stone (**a**), basalt (**b**) and magnetite (**c**) specimens adopted during the fatigue tests. The specimens present cylindrical shape, the diameter is 50 mm and are 100 mm high. All the specimens are drilled from blocks of querries located in Piedmont, Sardinia (magnetite) and Mount Etna in Sicily (basalt)

Table 4.1 Tested materials

Materials	Number of specimens	D (mm)	H (mm)	$\lambda = H/D$	Compressive strength (MPa)	Elastic modulus
Granitic orthogneiss	2	50	100	2	162 (MPa)	50 (MPa)
Basalt	2	50	100	2	170 (MPa)	60 (MPa)
Magnetite	2	50	100	2	250 (MPa)	220 (GPa)

4.2.2 Testing Procedure and Devices

For the lower frequency (2 Hz), six specimens, two for each kind of rock (D = 50 mm, H = 100 mm, $\lambda = 2$) were used. The tests were carried out by mean of MTS servo-controlled hydraulic testing machine with a maximum capacity of 1000 kN, working by a digital type electronic control unit (see Fig. 4.2a). The applied force was determined by measuring the pressure in the loading cylinder by means of a transducer. The margin of error in the determination of the force is 1 % which makes it a class 1 mechanical press.

The cyclic loading was programmed at a frequency of 2 Hz and with a load excursion from a minimum load of 10 kN to a maximum of 60 kN. With respect to the tests performed under monotonic displacement control, neutron emissions from compression tests under cyclic loading were measured by using neutron bubble detectors. Due to their isotropic angular response, three BDT and three BD-PND

Fig. 4.2 Servo-hydraulic press employed during the low frequency fatigue tests (**a**) Electrome-chanical Amsler Vibrofore 10 HPF 422 used during the experimental test characterized by a frequency working range between 100 and 260 Hz (**b**). Bandelin HD 2200 sonotrode used during ultrasonic tests (**c**)

detectors were positioned at a distance of about 5 cm all around the specimen. The detectors were previously activated, unscrewing the protection cap, to reach the suitable thermal equilibrium, and they were kept active throughout the test duration. Furthermore, a BDT and a BD-PND detector were used as background control during the test. Similar test configurations were adopted for basalt and magnetite specimens.

As far as the intermediate frequency (200 Hz) is concerned, the cylindrical specimens have been subjected to compression-compression fatigue tests by means of an electromechanical Amsler Vibrofore 10 HPF 422 (see Fig. 4.2b). This equipment allows to apply fatigue loading with a frequency range between 100 and 260 Hz, a mean load up to 50 kN, and an alternate load up to 50 kN. The machine working conditions are related to resonance conditions of the system which is composed by two masses (a seismic huge mass and the specimen mass) and two springs (a huge machine spring element and the specimen stiffness) disposed in series (see Fig. 4.2b). Working condition difficulties are often related to specimen damping. In the case of rocks, a tuning activity was needed to reach resonance conditions. The specimens were set between two compression platens, a preload (mean) was set and then alternate loading was applied.

Also in this case, the neutron emission was detected by using neutron bubble detectors. After 5×10^7 cycles the tests were interrupted. During the fatigue tests, two sets of measurements were performed: the specimen surface thermal infrared measurements and the neutron emission measurements by thermodynamic detectors.

The ultrasonic tests were conducted at a frequency of 20 KHz. Ultrasonic oscillation was generated by a high-intensity ultrasonic horn (Bandelin HD 2200) (see Fig. 4.2c). The device guarantees constant amplitude independently of changing conditions within the sample. The apparatus consists of a generator that converts electrical energy to 20 kHz ultrasounds, and of a transducer that switches this energy into mechanical longitudinal vibration at the same frequency. The specimens were connected to the ultrasonic horn by a glued screw inserted in a 5-mm deep hole (Fig. 4.2). This kind of connection was realized to achieve a resonance condition, considering the speed of sound in Luserna stone, and the length of the specimen. Ultrasonic irradiation of the specimen was carried out for 3 h. After the switching on of the transducer, 10 % of the maximum power was reached in 20 min. Successively, the transducer power increased to 20 % after 1 h, and then reached a maximum level of about 30 % after 2 h. Afterwards, the transducer worked at the same power condition up to the end of the test.

4.3 Neutron and Temperature Measurements

The thermal emission was acquired by means of a NEC TH7100WX infrared thermo camera (see Figs. 4.3a and 4.4). The specimens were black painted to maximize thermal emission. Thermal acquisition frame was set to 2 frame/min. The thermal acquisition system acquires the thermal contour of the element's surface set in front of the IR lens. In particular, the camera was focused on the specimen cylinder surface, as the radiation of the surface is perpendicular to the same surface, then the correct temperature measurement is obtained if the camera is set parallel to cylinder tangent surface. To avoid measurement errors, only the maximum measured temperatures on the specimens were recorded for the data processing. It can be verified that the maximum temperature is measured in correspondence of the line which lies in the tangent plane which is parallel to the

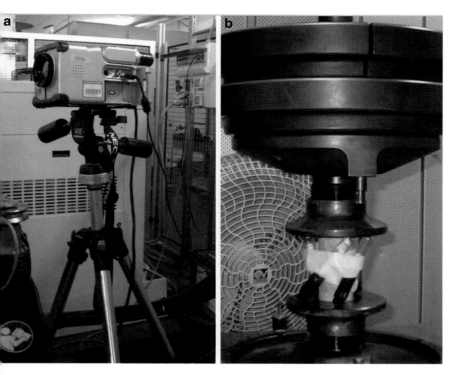

Fig. 4.3 NEC TH7100WX infrared thermo camera used for the temperature monitoring of the external surface of the specimen (**a**). A set of passive neutron detectors insensitive to electromagnetic noise was used for neutron emission detection (**b**)

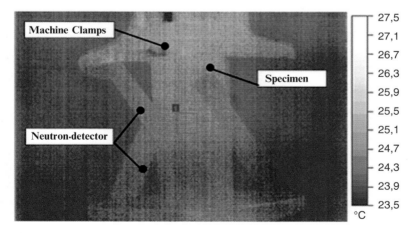

Fig. 4.4 Thermal contour of basalt specimen during compression fatigue loading. The specimen, the neutron dosimeters and the machine clamps are indicated. The detected temperature ranges from 23.5 °C to 27.5 °C

lens. The surface temperature values were processed by subtracting room temperature thus obtaining the surface thermal increment ΔT. It is well-known [17–20] that fatigue cyclic loading causes a thermal increment in rubber, ceramics, and metals. Different causes contribute [20, 21] to this phenomenon in the elastic conditions, the principal being thermoelastic effects and internal friction dissipations. The last phenomenon consists in a structural crystal re-organization implying a change in structural damping and in vibration frequencies, as well as microplastic deformation [20–25].

The neutron activity was detected by a thermodynamic Neutron Detection Technique used during all the fatigue experiments. Since neutrons are electrically neutral particles, they cannot directly produce ionization in a detector, and therefore cannot be directly detected. This means that neutron detectors must rely upon a conversion process where an incident neutron interacts with a nucleus to produce a secondary charged particle. These charged particles are then detected, and from them the neutron's presence is deduced. For an accurate neutron rate evaluation, a set of passive neutron detectors, based on the superheated bubble detection technique, insensitive to electromagnetic noise and with zero gamma sensitivity was employed (Fig. 4.3b). The dosimeters, based on superheated bubble detectors (BTI ON, Canada) (Bubble Technology Industries 1992), are calibrated at the factory against an AmBe (Americium-Beryllium) source in terms of NCRP38 (National Council on Radiation Protection and Measurements 1971). Bubble detectors are the most sensitive, accurate neutron dosimeters available that provide instant visible detection and measurement of neutron dose. Each detector is composed of a polycarbonate vial filled with elastic tissue equivalent polymer, in which droplets of a superheated gas (Freon) are dispersed. When a neutron strikes a droplet, the latter immediately vaporizes, forming a visible gas bubble trapped in the gel. The number of droplets provides a direct measurement of the equivalent neutron dose with an efficiency of about 20 %. These detectors are suitable for neutron dose measurements in the energy range of thermal neutrons (E = 0.025 eV, BDT type) and fast neutrons (E = 100 keV, BD-PND type).

4.4 Experimental Results

The preliminary experimental results reported in the literature [1–8] are confirmed by those obtained from the fatigue tests carried out on the cylindrical specimens. In addition, the experimental results seem to demonstrate that neutron emissions follow an anisotropic and impulsive distribution from a specific zone of the specimen. It is a matter of fact that the detected neutron flux, and consequently neutron dose, is inversely proportional to the square of the distance from the source. For these reasons, to avoid underestimated data acquisition, more bubble dosimeters placed around the test specimen were employed. For the different materials, droplets counting was performed every 12 h and the equivalent neutron dose was calculated, during low frequency fatigue experiments. In the same way, the natural

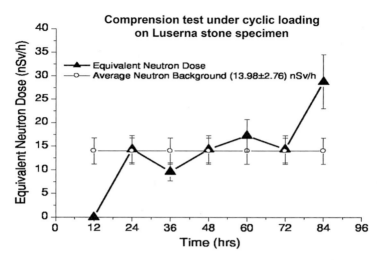

Fig. 4.5 Compression test under cyclic loading. Equivalent neutron dose variation on Luserna Stone specimen

background was estimated by means of the two bubble dosimeters used for assessment. For all the tests, the ordinary background was found to be (13.98 ± 2.76) nSv/h. In Fig. 4.5, equivalent neutron dose variation, evaluated during the cyclic (2 Hz) compression test of granitic orthogneiss, is reported. An increment of more than twice the background level was detected at the specimen failure. No significant variations in neutron emissions were observed before failure. The equivalent neutron dose, at the end of the test, was (28.74 ± 5.75) nSv/h.

As previously done, in Fig. 4.6a, b the results obtained by thermodynamic detectors and the equivalent neutron dose for a basalt specimen subjected to fatigue test at low frequency are reported. Also in this case, droplets counting was performed every 12 h and the equivalent neutron dose was calculated. During the test, the ordinary background was found to be (53.76 ± 13.44) nSv/h. The neutron equivalent dose variation, evaluated during the cyclic loading test, is reported (see Fig. 4.6b). An increment of about 20 times with respect to the background level was detected at specimen failure. No significant variations in neutron emissions were observed before failure. The equivalent neutron dose, at the end of the test, was (935.49 ± 233.87) nSv/h.

As far as the magnetite specimen is concerned, the ordinary background was found to be (15.05 ± 3.76) nSv/h. For this very iron-rich rock the neutron equivalent dose variation, evaluated during the cyclic loading test at low frequency (2 Hz), is reported in Fig. 4.7. An increment of about 100 times with respect to the background level was detected at failure. No significant variations in neutron emissions were observed before failure. The equivalent neutron dose, at the end of the test, was (1036.78 ± 259.19) nSv/h.

In Fig. 4.8, the neutron emission level is reported for a comparison between granitic orthogneiss, basalt, and magnetite, subjected to fatigue experiments with

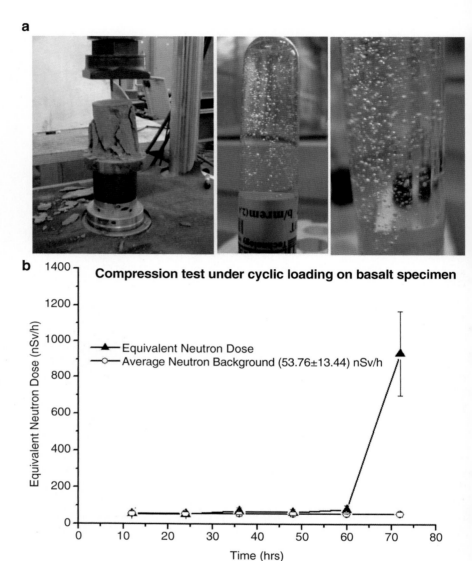

Fig. 4.6 Results from bubble thermodynamic dosimeters obtained at the end of the test considering low frequency (2 Hz) on basalt (**a**). Equivalent neutron dose variation during the test (**b**)

intermediate loading frequency (200 Hz). It is interesting to note that the emission trends are conserved considering the behaviour obtained in the case of low frequency. Also in this case, in fact, the granitic specimens show a small but appreciable increase. The basalt specimens show an increase up to one order of magnitude greater than the background level, and the magnetite specimens show a neutron emission up to two orders of magnitude greater than the background level.

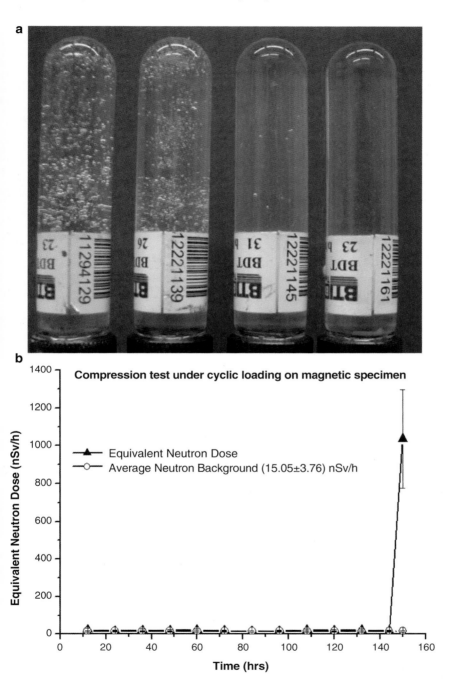

Fig. 4.7 Results from bubble thermodynamic dosimeters obtained at the end of the test considering low frequency (2 Hz) on Magnetite (a). Equivalent neutron dose variation during the test (b)

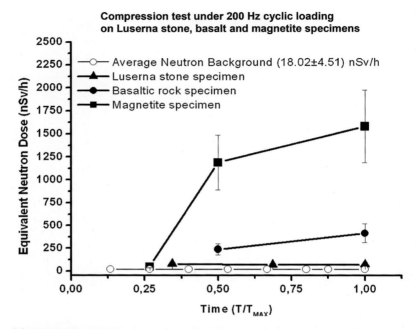

Fig. 4.8 Results from bubble thermodynamic dosimeters obtained at the end of the test considering intermediate frequency (200 Hz) on granite, basalt and magnetite

The last kind of experiments were conducted using ultrasound vibration applied to the specimens. The ultrasonic test on the Luserna stone specimen (D = 53 mm, H = 100 mm) was carried out at the Medical and Environmental Physics Laboratory of the Experimental Physics Department of the University of Turin. A relative natural background measurement was performed by means of the He^3 detector for more than 6 h. The average natural background was $(6.50 \pm 0.85) \times 10^{-3}$ cps, for a corresponding thermal neutron flux of $(1.00 \pm 0.13) \times 10^{-4}$ $n_{thermal}$ cm^{-2} s^{-1}. This natural background level, lower than the one calculated during the monotonic compression tests at the Fracture Mechanics Laboratory of the Politecnico of Torino, is related to the location of the Experimental Physics Laboratory, which is three floors below the ground level. Similar experiments were conducted using a basalt specimen. In this case, the thermal neutron flux was about two orders of magnitude greater than the background level. Ultrasonic experiments were not conducted on magnetite due to the brittleness of this rock. For the magnetite, in fact, the connection to the ultrasonic horn by a glued screw inserted in a 5-mm deep hole was not possible, due to the sudden failure of the specimen after few minutes from the beginning of the test.

During the ultrasonic test, the specimen temperature was monitored by using a multimeter/thermometer (Tektronix mod. S3910). The temperature reached 50 °C after 20 min, and then increased up to a maximum level of 100 °C at the end of the ultrasonic test. In Fig. 4.9, the neutron emissions detected are compared with the

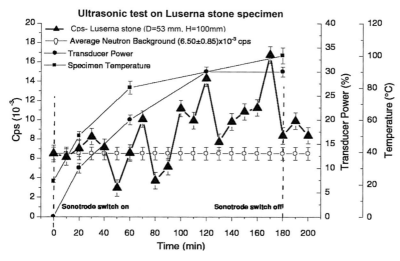

Fig. 4.9 Results from bubble thermodynamic dosimeters obtained at the end of the test considering ultrasonic frequency (20 kHz) on Luserna stone

transducer power trend and with the specimen temperature. A significant increment in neutron activity after 130 min from the beginning of the test was measured (see Fig. 4.9). At this time, the transducer power reached 30 % of the maximum, with a specimen temperature of about 90 °C. Some neutron variations were detected during the first hour of the test, but they may be due to ordinary fluctuations of natural background. At the switching off of the sonotrode, the neutron activity decreased to the typical background value.

In Fig. 4.10, the results of neutron emission to neutron background ratio for the different kinds of test and rock are reported in order to put in comparison the neutron flux, the frequency range, and the iron content in the tested specimens. To evaluate all the possible cases, the neutron emissions were considered also in the case of monotonic loading condition, assuming the results obtained from the static tests reported by Carpinteri et al. [3–8]. From the results shown in Fig. 4.10, it is possible to observe that the maximum neutron emission level starts to decrease from monotonic loading condition, in which the neutron emission seems to be maximized for all the materials, down to the ultrasonic range (20 kHz), in which the neutron emissions for Luserna stone and basalt is only some times greater than the background level detected before the experiments. Particular consideration has to be given to the case of fatigue loading at the intermediate frequency (200 Hz). For this frequency range, between 100 and 260 Hz, an higher value for the neutron emission to neutron background ratio can be observed for all the tested materials (see Fig. 4.10). This frequency seems to generate the highest neutron emissions. It is also interesting to consider that the frequency range around 200 Hz, for which the neutron emission is maximized, can be recorded in the preparation zone of the earthquakes. In particular, this kind of micro-seismic activity were observed in

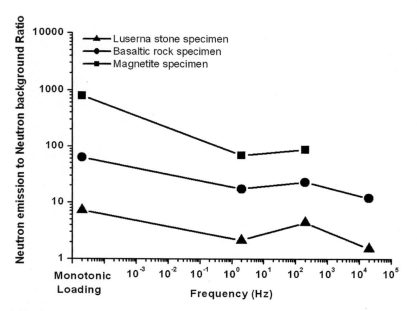

Fig. 4.10 Comparison between the different neutron emission to neutron background ratio for monotonic loading condition and fatigue conditions ranging from 2 Hz to 20 kHz

recent monitoring of seismic signal detection in correspondence of the Jacinto fault propagation (Southern California) [26]. At the same time, in the preparation zone of an impending earthquake, strain in collapsing pores, grain boundary slippage, and micro-fractures may cause acoustic emissions in a frequency range up to 300 kHz [27–29]. These results may be correlated to the neutron emissions recently observed in correspondence of earthquakes, leading to consider also the Earth's Crust, in addition to cosmic rays, as being a relevant source of neutron flux variations [30–34]. Neutron emissions exceeded the neutron background up to 1000 times in correspondence to seismic events with a Richter magnitude equal to the 4th degree [5, 30–34]. From this point of view, it is interesting to note that, from the laboratory to the Earth' Crust scale, the frequency of 200 Hz, produced by the testing machine or by the micro-seismic activity, may be considered as a catalyzing mechanism able to maximize the neutron emissions during cyclic solicitations.

Comparing the thermal increment for the three different investigated materials, only basalt shows relevant increases. The thermal increment seems to be related both to mean load and to alternate load. It can also be noted that subsequent loading blocks with the same loading level show increasing thermal increments (see Fig. 4.11). This behaviour is consistent with metals and ceramics thermal behaviour in fatigue loading [22, 23]. In these cases, in fact, the higher the temperature the higher the damage level localized in the specimen. When a specimen is already damaged, the subsequent damage is more effective, and so it generates a higher dissipation and a thermal increase. It has to be noted that the damage process occurs

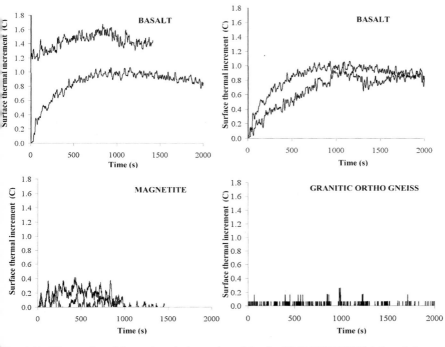

Fig. 4.11 The results of thermal analysis conducted by the NEC TH7100WX infrared thermo camera have been reported. For each graph the results of two tests are shown

during testing, even for loading values which are very low if compared to the mechanical properties of the specimens. According to the literature [20, 24], fatigue damage occurs for any loading value, even below the fatigue limit, and is related to microplastic damaging in microsites which are statistically activated, depending on the loading value. In any case, a structural modification in the material occurs and it can be detected in equivalent neutron dose variations and thermal increments detected during the tests.

4.5 Conclusions

Fatigue experiments at low (2 Hz), intermediate (200 Hz), and high (20 kHz) frequency are performed with different iron-bearing materials. The results confirm that neutron emissions greater than the background level may be observed during damage accumulation in granitic orthogneiss, basalt, and magnetite. The neutron detection, together with temperature measurements obtained by infrared revelation, lead to the conclusion that fatigue tests performed at 200 Hz represent the condition for which the neutron emission is the highest for specimens subjected to cyclic loading.

As summarized in Fig. 4.10, the emission of neutrons can be related to the iron content in the different materials, and to the testing condition used during the experiments (monotonic or cyclic at different working frequencies). It can be observed that the higher the iron content, the greater the neutron emission. The high neutron flux emitted from magnetite during brittle fracture is confirmed in the case of fatigue tests performed at the intermediate frequency (200 Hz). The Etna basalt shows an average emission about one order of magnitude greater than those found in Luserna stone. Etna basalt shows, at the same time, the most relevant thermal increment measured by infrared waves (see Fig. 4.11).

As regards granite in the different testing conditions, it is possible to confirm that the neutron emission is equal to several times the background level for cyclic loads, and up to ten times the background in the case of specimen crushing.

Particular attention has to be drawn in the case of fatigue at the intermediate frequency of 200 Hz. At this frequency an higher value of the neutron emission to neutron background ratio can be observed for all the tested materials (see Fig. 4.10). This frequency seems to generate higher neutron emissions than the other loading frequencies investigated. It is also interesting to consider that the frequency range around 200 Hz, for which the neutron emission is maximized, can be also recorded in the preparation zone of earthquakes and during fault propagation. In particular, this kind of micro-seismic activity was observed in the recent monitoring carried out in correspondence to the Jacinto fault propagation in Southern California. At the same time, in the preparation zone of an impending earthquake, strain in collapsing pores, grain boundary slippage, and micro-fractures may cause acoustic emissions in a frequency range up to 300 kHz [26–29]. These results may be correlated to the neutron emissions recently observed in correspondence to earthquakes, leading to consider also the Earth's Crust, in addition to cosmic rays, as being a relevant source of neutron flux variations [30–34]. Neutron emissions exceeded the neutron background up to 1000 times in correspondence to seismic events with a Richter magnitude equal to the 4th degree [5, 30–34].

References

1. Carpinteri A, Cardone F, Lacidogna G (2009) Piezonuclear neutrons from brittle fracture: early results of mechanical compression tests. Strain 45:332–339
2. Cardone F, Carpinteri A, Lacidogna G (2009) Piezonuclear neutrons from fracturing of inert solids. Phys Lett A 373:4158–4163
3. Carpinteri A, Cardone F, Lacidogna G (2010) Energy emissions from failure phenomena: mechanical, electromagnetic, nuclear. Exp Mech 50:1235–1243
4. Carpinteri A, Borla O, Lacidogna G, Manuello A (2010) Neutron emissions in brittle rocks during compression tests: monotonic vs. cyclic loading. Phys Mesomech 13:268–274

5. Carpinteri A, Lacidogna G, Manuello A, Borla O (2012) Piezonuclear fission reactions: evidences from microchemical analysis, neutron emission, and geological transformation. Rock Mech Rock Eng 45:445–459
6. Carpinteri A, Lacidogna G, Manuello A, Borla O (2013) Piezonuclear fission reactions from earthquakes and brittle rocks failure: evidence of neutron emission and nonradioactive product elements. Exp Mech 53(3):345–365
7. Carpinteri A, Lacidogna G, Manuello A, Borla O (2013) Energy emissions from brittle fracture: neutron measurements and geological evidences of piezonuclear reactions. Strength Fract Complex 7:13–31
8. Carpinteri A, Lacidogna G, Borla O, Manuello A, Niccolini G (2012) Electromagnetic and neutron emissions from brittle rocks failure: experimental evidence and geological implications. Sadhana 37:59–78
9. Widom A, Swain J, Srivastava YN (2015) Photo-disintegration of the iron nucleus in fractured magnetite rocks with magnetostriction. Meccanica 50:1205–1216
10. Widom A, Swain J, Srivastava YN (2013) Neutron production from the fracture of piezoelectric rocks. J Phys G Nucl Part Phys 40:15006 (1–8)
11. Vola G, Marchi M (2010) Quanitative phase analysis (QPA) of the Luserna stone. Period Miner 79(2):45–60
12. Sandrone R, Cadoppi P, Sacchi R, Vialon P (1993) The Dora-Maira Massif. In: Von Raumer F, Neubauer F (eds) Pre-mesozoic geology in the alps. Springer, Berlin, pp 317–325
13. Sandrone R, Borghi A (1992) Zoned garnets in the northern Dora-Maria Massif and their contribution to a reconstruction of the regional metamorphic evolution. Eur J Mineral 4:465–474
14. Sandrone R, Colombo A, Fiora L, Fornaro M, Lovera E, Tunesi A, Cavallo A (2004) Contemporary natural stones from the Italian western Alps (Piedmont and Aosta Valley regions) period. Mineral 73:211–226
15. Tanguy JC (1978) Tholeiitic basalt magmatism of Mount Etna and its relations with the alkaline series. Contrib Mineral Petrol 66:51–67
16. Verkaeren J, Bartholomè P (1979) Petrology of the San Leone magnetite skarn deposit (S. W. Sardinia). Econ Geol 74:53–66
17. Chrysochoos A, Louche H (2000) An infrared image processing to analyse the calorific effects accompanying strain localisation. Int J Eng Sci 28:1759–1788
18. Curà F, Gallinatti AE, Sesana R (2012) Dissipative aspects in thermographic methods. Fatigue Fract Eng Mater Struct 5:1133–1147
19. Kim J, Jeong HY (2010) A study on the hysteresis, surface temperature change and fatigue life of SM490A, SM490A-weld and FC250 metal materials. Int J Fatigue 32:1159–1166
20. Crupi V (2008) An unifying approach to assess the structural strength. Int J Fatigue 30:1150–1159
21. Luong MP (1998) Fatigue limit evaluation of metals using an infrared thermographic technique. Mech Mater 28:155–163
22. Curà F, Curti G, Gallinatti AE, Sesana R (2006) Modello termomeccanico per l'analisi del danneggiamento in provini sottoposti a fatica. In: Proceedings of XXXV Conference AIAS, Ancona, Italy, 13–16 September
23. Curti G, Curà F, Sesana R (2006) Thermomechanical model and experimental analysis of progressive fatigue damage in steels specimens. In: Proceedings of ESDA 2006, 8th Biennial ASME Conference, Turin, Italy, 4–7 July
24. Doudard C, Calloch S, Hild F, Cugy P, Galtier A (2005) A probabilistic two scale model for high cycle fatigue life predictions. Fatigue Fract Eng Mater Struct 28:279–288
25. Curti G, Curà F, Sesana R (2005) A new iteration method for the thermographic determination of fatigue limit in steels. Int J Fatigue 27:453–459
26. Aster RC, Shearer PM (1991) High-frequency borehole seismograms recorded in the San Jacinto Fault zone, Southern California part 2 attenuation and site effects. Bull Seismol Soc Am 81(4):1081–1100

27. Lockner DA, Byerlee JD, Kuksenko V, Ponomarev A, Sidorin A (1991) Quasi-static fault growth and shear fracture energy in granite. Nature 350:39–42
28. Rabinovitch A, Frid V, Bahat D (2007) Surface oscillations – a possible source of fracture induced electromagnetic radiation. Technophysics 431:15–21
29. Amstrong BH, Valdes CM (1991) Acoustic emission/microseismic activity at very low strain levels. In: Sachse W, Roget J, Yamaguchi K (eds) Acoustic emission: current practice and future directions, ASTM STP 1077. American Society for Testing and Materials, Philadelphia
30. Kuzhevskij BM, Nechaev OY, Sigaeva EA, Zakharov VA (2003) Neutron flux variations near the Earth's crust. A possible tectonic activity detection. Nat Hazards Earth Syst Sci 3:637–645
31. Kuzhevskij BM, Nechaev OY, Sigaeva EA (2003) Distribution of neutrons near the Earth's surface. Nat Hazards Earth Syst Sci 3:255–262
32. Antonova VP, Volodichev NN, Kryukov SV, Chubenko AP, Shchepetov AL (2009) Results of detecting thermal neutrons at Tien Shan high altitude station. Geomagn Aeron 49:761–767
33. Volodichev NN, Kuzhevskij BM, Nechaev OY, Panasyuk MI, Podorolsky AN, Shavrin PI (2000) Sun-Moon-Earth connections: the neutron intensity splashes and seismic activity. Astron Vestn 34:188–190
34. Sigaeva E et al (2006) Thermal neutrons'observations before the Sumatra earthquake. Geophys Res Abstr 8:00435

Chapter 5
Alpha Particle Emissions from Carrara Marble Specimens Crushed in Compression and X-ray Photoelectron Spectroscopy of Correlated Nuclear Transmutations

Alberto Carpinteri, Giuseppe Lacidogna, and Oscar Borla

Abstract Neutron emission measurements were carried out by means of a He^3 detector on Carrara marble specimens under compression. While granite generated neutrons − due to piezonuclear reactions involving fission of iron into aluminum − this phenomenon did not appear in marble crushing tests. On the other hand, significant alpha particle fluctuations were detected by a 6150 AD-k probe during the same compression tests.

The external and fracture surfaces belonging to Carrara marble specimens crushed during the compression tests were analyzed by X-ray Photoelectron Spectroscopy (XPS). Such quantitative compositional analyses were carried out in order to detect any variation in Carrara marble chemical composition due to brittle failure. A total decrement in Ca, Mg, and O by 13 % as well as an equivalent increment in C were observed on the fracture surface with respect to the external surface. The assumed transmutations involve elements with an equal number of protons and neutrons. For this reason, the micro-chemical analyses suggest piezonuclear reactions accompanied by alpha particle emissions, but without neutron emissions, in the crushing experiments on marble specimens.

Keywords X-ray photoelectron spectroscopy • Alpha particle emissions • Nuclear transmutations • Carrara marble

5.1 Introduction

After summarizing the preliminary results already presented in [1–3], involving compression tests on Luserna stone, we present new experiments performed on Carrara marble specimens under mechanical loading. The neutron background

A. Carpinteri (✉) • G. Lacidogna • O. Borla
Department of Structural, Geotechnical and Building Engineering,
Politecnico di Torino, Corso Duca degli Abruzzi 24, 10129 Torino, Italy
e-mail: alberto.carpinteri@polito.it

© Springer International Publishing Switzerland 2015
A. Carpinteri et al. (eds.), *Acoustic, Electromagnetic, Neutron Emissions from Fracture and Earthquakes*, DOI 10.1007/978-3-319-16955-2_5

monitoring, by means of He^3 devices and bubble detectors, was performed during the compression tests under monotonic displacement control. The compression tests on specimens with different sizes and shapes were carried out at the Fracture Mechanics Laboratory of the Politecnico di Torino.

In the papers [1–3] we found that neutron emissions from Luserna stone specimens of larger dimensions exceeded the background level by approximately one order of magnitude at failure. In the compression of Carrara marble specimens the neutron measurements yielded values comparable to the background level, even at the time of failure. While Luserna stone generated neutrons – due to piezonuclear reactions involving fission of iron into aluminum – this phenomenon did not appear in marble crushing. However, even if no relevant neutron emissions were detected during mechanical loading of marble, important piezonuclear reactions have been observed, substantiated by significant alpha particles flux variations, during the compression tests. Moreover, similar evidence of alpha particles emission has been recently observed during cyclic loading tests on cylindrical steel bars [4].

The external and the fracture surfaces belonging to Carrara marble specimens crushed during the compression tests were analyzed by X-ray Photoelectron Spectroscopy (XPS). A decrement in Ca, Mg, and O as well as an increment in C contents were observed on the fracture surfaces with respect to the external surfaces. These analyses suggest piezonuclear reactions with alpha particle emissions and without neutron emissions.

A theoretical explanation about piezonuclear fission reactions has been recently provided based on classical Nuclear Physics [5, 6].

5.2 Neutron and Alpha Particle Emission Detection Techniques

Since neutrons are electrically neutral particles, they cannot directly produce ionization in a detector, and therefore cannot be directly detected. This means that neutron detectors must rely upon a conversion process where an incident neutron interacts with a nucleus to produce a secondary charged particle. These charged particles are then detected, and from them the neutrons presence i deduced. For an accurate neutron evaluation, a He^3 proportional was used.

Moreover, a set of passive neutron detectors, based on the superheated bubble detection technique and insensitive to electromagnetic noise, were employed.

5.2.1 He^3 Neutron Proportional Counter

The He^3 detector used in the compression tests under monotonic displacement control and by ultrasonic vibration, is a He^3 type (Xeram, France) with electronics of preamplification, amplification, and discrimination directly connected to the detector tube. The detector, filled with 4 bars of helium-3 gas, is powered with 1.3 kV, supplied via a high voltage NIM (Nuclear Instrument Module). The logic output producing the TTL (transistor-transistor logic) pulses is connected to a NIM counter. The device was calibrated for the measurement of thermal neutrons; its sensitivity is 65 cps/n thermal ($\pm$ 10 % declared by the factory), i.e., a thermal neutron flux of 1 thermal neutron/s cm^2 corresponds to a count rate of 65 cps.

5.2.2 Neutron Bubble Detectors

A set of passive neutron detectors insensitive to electromagnetic noise and with zero gamma sensitivity was used in compression tests under cyclic loading. The dosimeters, based on superheated bubble detectors (BTI, Ontario, Canada) (Bubble Technology Industries (1992)) [7], are calibrated at the factory against an Am-Be source in terms of NCRP38 (NCRP report 38 (1971)) [8]. Bubble detectors provide instant visible detection and measurement of neutron dose. Each detector is composed of a polycarbonate vial filled with elastic tissue-equivalent polymer, in which droplets of a superheated gas (Freon) are dispersed. When a neutron strikes a droplet, the latter immediately vaporizes, forming a visible gas bubble trapped in the gel. The number of droplets provides a direct measurement of the equivalent neutron dose. These detectors are suitable for neutron integral dose measurements, in the energy ranges of thermal neutrons (E = 0.025 eV) and fast neutrons (E > 100 keV).

5.2.3 The 6150 AD-k Probe

For the alpha particle emissions, a 6150 AD-k probe with a sealed proportional counter was used. A peculiarity of this device is that it does not require refilling or flushing from external gas reservoirs. The probe is sensitive to alpha and/or beta, and gamma radiation by means of a removable discriminator plate (stainless steel, 1 mm). An electronic switch allows for the operating mode "alpha" to detect alpha radiation only. In this operational mode, the radiation detection is very sensitive because the background level is much lower. During the experiments, the 6150 AD-k probe was used in the operating mode "alpha" to monitor the background level before and after the switching on of the cell.

5.3   Preliminary Tests on Luserna Stone Specimens

During previous experimental compression tests, neutron emissions were measured on nine Luserna stone cylindrical specimens, of different size and shape (Table 5.1), denoted with P1, P2,..., P9 [1–3]. In the following the main results are briefly presented. The He^3 neutron detector was switched on at least 1 h before the beginning of each compression test, in order to reach the thermal equilibrium of electronics, and to make sure that the behaviour of the device was stable with respect to intrinsic thermal effects. The detector was placed in front of the test specimen at a distance of 10 cm and it was enclosed in a polystyrene case in order to avoid "spurious" signals coming from impacts and vibrations.

Neutron measurements for specimens P2, P3, P4, P7 yielded values comparable with the ordinary natural background, whereas in specimens P1 and P5 the experimental data exceeded the background level by approximately four times. For specimens P6 neutron emissions of about five times the background level were observed concomitant with the sharp stress drop at the time of failure, while for specimens P8 and P9 the neutron emissions achieved values by approximately one order of magnitude higher than the ordinary background. In Table 5.1, the experimental data concerning compression tests on the nine Luserna stone specimens are summarized. These phenomena were assumed to be caused by piezonuclear reactions occurring for iron atoms [9, 10].

Considering the experimental results obtained by the authors in a previous work [11], a volume approximately exceeding 200,000 mm^3, combined with the extreme brittleness of the tested material, represents a threshold value for a neutron emission of about one order of magnitude higher than the ordinary background. As regards the expected energy spectrum, it extends from thermal neutrons (0.025 eV) up to the fast component (few MeV). Also this behaviour has already been measured by the authors [1–3] by using specific devices such as proportional counters (He^3 devices) and passive bubble dosimeters. In the future, more detailed information about the energy spectrum using special spectrometers will be provided.

5.4   Tests on Carrara Marble Specimens Experimental Set-Up and Results

Neutron emissions were measured on 27 Carrara marble cylindrical specimens, three for each size and shape (see Table 5.2), denoted with M1, M2,..., M9. In Fig. 5.1a some of the Carrara marble test specimens with the same size but different slenderness are shown. The tests were carried out by means of a MTS servohydraulic press, with a maximum capacity of 1000 kN, working by a digital type electronic control unit. The force applied was determined by measuring the pressure in the loading cylinder by means of a transducer. The specimens were arranged with the two smaller surfaces in contact with the press platens, without coupling

Table 5.1 Compression tests under monotonic displacement control. Specimen dimensions and neutron emissions from Luserna stone specimens

Specimens	Dimension Diameter [mm]	Slenderness λ	Piston velocity [m/s]	Volume [mm³]	Average peak load [kN]	Average neutron background (10^{-2} cps)	Count rate at the neutron emission (10^{-2} cps)
P1	28	0.5	1×10^{-6}	8,616	52.19	3.17 ± 0.32	8.33 ± 3.73
P2	28	1	1×10^{-6}	17,232	33.46	3.17 ± 0.32	Background
P3	28	2	1×10^{-6}	34,464	41.28	3.17 ± 0.32	Background
P4	53	0.5	1×10^{-6}	58,434	129.00	3.83 ± 0.37	Background
P5	53	1	1×10^{-6}	116,868	139.10	3.84 ± 0.37	11.67 ± 4.08
P6	53	2	1×10^{-6}	233,736	206.50	4.74 ± 0.46	25.00 ± 6.01
P7	112	0.5	1×10^{-5}	551,434	1099.30	4.20 ± 0.80	Background
P8	112	1	1×10^{-5}	1,102,868	1077.10	4.20 ± 0.80	30.00 ± 11.10
P9	112	2	1×10^{-5}	2,205,736	897.80	4.20 ± 0.80	30.00 ± 10.00

Table 5.2 Carrara marble compression tests under monotonic displacement control. Average specimen dimensions and neutron emissions

Specimens	Dimension Diameter [mm]	Slenderness λ	Piston velocity [m/s]	Volume [mm³]	Average peak load [kN]	Average neutron background (10^{-2} cps)	Count rate at the neutron emission (10^{-2} cps)
M1	25	0.5	5×10^{-7}	6,133	69.0	5.64 ± 1.41	Background
M2	25	1	1×10^{-6}	12,266	47.4	5.64 ± 1.41	Background
M3	25	2	2×10^{-6}	24,532	36.1	5.64 ± 1.41	Background
M4	50	0.5	5×10^{-7}	49,062	223.7	7.29 ± 1.83	Background
M5	50	1	1×10^{-6}	98,125	164.1	7.29 ± 1.83	Background
M6	50	2	2×10^{-6}	196,250	144.5	7.29 ± 1.83	Background
M7	100	0.5	5×10^{-7}	392,500	909.0	6.25 ± 1.55	Background
M8	100	1	1×10^{-6}	785,000	880.6	6.25 ± 1.55	Background
M9	100	2	2×10^{-6}	1,570,000	714.0	6.25 ± 1.55	Background

Fig. 5.1 (**a**) Carrara marble cylindrical specimens of constant size, by varying slenderness. (**b**) Carrara marble specimen M9 after compression fracture. The bubble detectors have been placed all around the monitored specimen. On the *left* it has been positioned the He^3 neutron detector at a distance of about 20 cm

materials in-between, according to the testing modalities known as "test by means of rigid platens with friction". The platens were controlled by means of a wire-type potentiometric displacement transducer. The tests were performed under displacement control, with the planned displacement velocity equal to 0.0005 mm/s for the specimens with a slenderness $\lambda = 0.5$, 0.001 mm/s for $\lambda = 1$, and 0.002 mm/s for $\lambda = 2$.

A relative measurement of natural neutron background was performed in order to assess the average background affecting data acquisition in experimental room condition. The He^3 device was positioned in the same condition of the experimental set up and the background measures were performed fixing at 60 s the acquisition time, during a preliminary period of more than 3 h, for a total number of 200 counts. The average measured background level is ranging from $(5.64 \pm 1.41) \times 10^{-2}$ to $(7.29 \pm 1.83) \times 10^{-2}$ cps (see Table 5.2).

Additional background measurements were repeated before each test, fixing an acquisition time of 60s in order to check a possible variation in natural background. Neutron measurements for all specimens yielded values comparable to the ordinary natural background. In Table 5.2, the experimental data concerning compression tests on the Carrara marble specimens are summarized. In Fig. 5.1b, Carrara marble specimen M9 is shown after compression fracture.

As an example, in Fig. 5.2 the load vs. time diagram, and the neutron count rate evolution for specimens M3, M6, and M9 are shown. We can observe that the specimens of smaller dimension (M3) present a ductile behaviour in compression, whereas, by increasing the sample size, the behaviour becomes more brittle. As a matter of fact, specimens M9 show a catastrophic failure [1].

Droplets counting for the bubble detectors was performed at the end of each compression test and the equivalent neutron dose was calculated. In the same way,

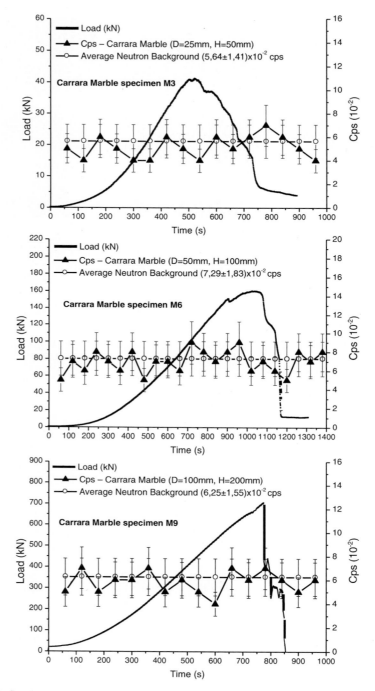

Fig. 5.2 Specimens M3, M6, M9. Load versus time diagrams, and neutron emission count rate

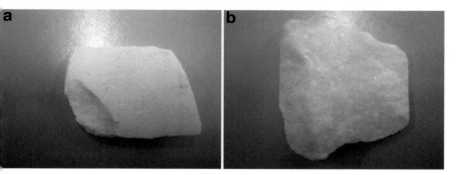

Fig. 5.3 (a) Portion of an external specimen surface. (b) Fracture surface belonging to the same specimen

the natural background was estimated by means of two bubble dosimeters used for assessment. The ordinary background was found to be (20.45 ± 4.09) nSv/h. Similarly to the He^3 results, no relevant neutron emissions were detected.

5.5   Compositional and Microchemical Evidence of Piezonuclear Fission Reactions in the Rock Specimens

X-ray Photoelectron Spectroscopy (XPS) was performed on different samples of external and fracture surfaces belonging to the three M9 marble specimens crushed during the compression tests. The measurement precision is in the order of magnitude of 0.1 %. XPS quantitative compositional analyses were carried out in order to detect any variation in Carrara marble composition. The XPS analysis was developed with a VersaProbe5,000 Physical Electronics instrument, a monochromatic Al source (1486.6 eV) and a hemispherical analyzer.

Survey scans as well as narrow scans were recorded with a 100 μm spot depth on an area of (500×500) μm^2. An electron flood gun combined with Argon ion gun was used to control surface charging during the measurements. XPS survey scan (pass energy of 187.85 eV) of Carrara marble surfaces showed that Ca, C, Mg and O were present. The high-resolution C(1s), Ca(2p), Mg(1s) and O(1s) spectra were obtained at a pass energy of 23.50 eV.

In Fig. 5.3a, the external surface of a portion of one of the three M9 specimens is shown. In Fig. 5.3b, the fracture surface of a fragment taken from the same specimen is shown. For the XPS analyses, several sites were localized on the surface of the thin sections and on the fracture surfaces. Thirty spots on the external surface and twenty on the fracture surface were selected and analysed.

In Figs 5.4a–d, the results for the Ca, Mg, O, and C concentrations are shown.

It can be observed that the distributions of O, Ca and Mg for the external surface, represented in the graphs by squares, show an average value (calculated as the

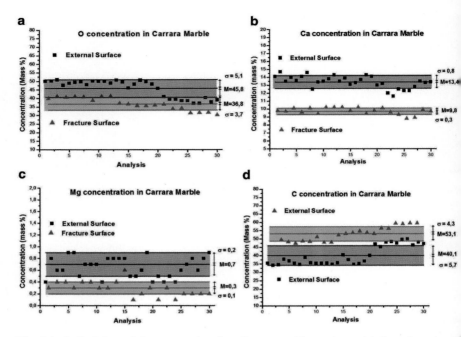

Fig. 5.4 O, Ca, Mg, and C concentrations in a Carrara marble specimen: (**a**) O on the external surface (*squares*) and on the fracture surface (*triangles*). (**b**) Ca on the external surface (*squares*) and on the fracture surface (*triangles*). (**c**) Mg on the external surface (*squares*) and on the fracture surface (*triangles*). (**d**) C on the external surface (*squares*) and on the fracture surface (*triangles*)

arithmetic mean value) respectively equal to 45.8 %, 13.4 % and 0.7 %. In the same graphs, the corresponding distributions of O, Ca and Mg concentrations on the fracture surface (indicated by triangles) show significant variations. It can be seen that the mean values in this case are respectively equal to 36.8 %, 9.8 % and 0.3 %, and they are considerably lower than the mean values related to the external surface measurements.

Similarly to Figs. 5.4a–c, in Fig. 5.4d the C mass percentage concentrations are considered in both the cases of external and fracture surfaces. The observed variations show a mass percentage increase approximately equal to the percentage decrease in O, Ca, and Mg. The average increase in the distribution corresponding to the fracture surface (indicated by triangles), is approximately 13.0 %. The average value of C concentrations changes from 40.1 % on the external surface to 53.1 % on the fracture surface. The evidence emerging from the XPS analyses, that the values of oxygen, calcium and magnesium decrease (−13 %) and that of C increase (+13.0 %) are equal, is really impressive (see Table 5.3).

From the results shown in the previous diagrams it can be clearly seen that piezonuclear reactions are possible in the inert non-radioactive Carrara marble specimens. From the XPS results on fracture samples, the evidences of O, Ca,

Table 5.3 Mass percentage concentrations in the external and fracture surfaces

| | Mass percentage concentrations | | |
	External surface	Fracture surface	Difference of mass percentage concentrations
Oxygen	(45.8 ± 5.1)	(36.8 ± 3.7)	-9.0
Calcium	(13.4 ± 0.8)	(9.8 ± 0.3)	-3.6
Magnesium	(0.7 ± 0.2)	(0.3 ± 0.1)	-0.4
Carbon	(40.1 ± 5.7)	(53.1 ± 4.3)	$+13.0$

Mg, and C variations lead to the conclusion that the following piezonuclear reactions should have occurred:

$$O_8^{16} \rightarrow C_6^{12} + He_2^4 \tag{5.1}$$

$$Ca_{20}^{40} \rightarrow 3C_6^{12} + He_2^4 \tag{5.2}$$

$$Mg_{12}^{24} \rightarrow 2C_6^{12} \tag{5.3}$$

These transmutations involve elements (O, Ca and Mg) with an equal number of protons and neutrons, like their reaction products (C and He). For this reason, the micro-chemical analyses suggest piezonuclear reactions without neutron emissions from marble specimens. In addition, further experimental confirmations on the assumption of piezonuclear reactions (5.1) and (5.2) have been observed during specific compression tests on Carrara marble samples monitored by an alpha particle detector.

The authors are also aware of the potential problem of hydrocarbon impurity, the so-called "adventitious carbon", that can reduce the interfacial energy during XPS analysis on calcium carbonate surfaces [12]. Moreover, Carrara marble is considered as a very homogeneous rock due to its relatively simple chemical composition and poor impurities. For this reason it is quite rare to find regions of higher carbon concentration that can be linked to weak surfaces susceptible to preferential fractures. To avoid incorrect Ca/C and Ca/O atomic ratios, further analyses are now in progress. These new results will be discussed in future publications.

5.6 Tests on Carrara Marble Specimens: Alpha Particle Monitoring

Alpha particle emissions were measured on three Carrara marble cylindrical specimens (A1, A2, and A3) with different slenderness, from 0.5 to 2. Similarly to the neutron monitoring, the tests were carried out by means of a MTS servo-hydraulic press, with a maximum capacity of 1000 kN. The tests were performed under displacement control, with the planned displacement velocity equal to

0.0005 mm/s for the specimen with a slenderness $\lambda = 0.5$, 0.001 mm/s for $\lambda = 1$, and 0.002 mm/s for $\lambda = 2$.

A relative measurement of alpha background was performed in order to assess the average background affecting data acquisition in experimental room condition. The 6150 AD-k probe was positioned in the same condition as the experimental set up during a preliminary period of 24 h, to detect the typical daily fluctuation of alpha field mostly due to radon concentration variability [13]. The average measured background level is approximately of $(12.00 \pm 3.00) \times 10^{-2}$ cps.

In Table 5.4, the experimental data concerning the compression tests on the Carrara marble specimens are summarized.

In Fig. 5.5 the load vs. time diagram and the alpha count rate evolution for specimen A3 is reported. The specimen shows a catastrophic behaviour at the final collapse with a maximum alpha emission of $(110.00 \pm 28.00) \times 10^{-2}$ cps, about ten times higher than the environmental level.

For all the specimens we observed a considerable increment in the alpha counting rate (counts per second) acquired during all the tests. The counting rate decreased to the typical background level at the end of each experiment.

On the alpha particles tracking the following further considerations can be highlighted. As experimentally observed by the authors in the present paper, as well as in other published articles, large-amplitude high-frequency vibrations produced during a brittle fracture process generate internal excitation of the nucleous inducing nuclear disintegration, elemental variations, low-frequency electromagnetic emission, neutron emission, and alpha particle emission. The occurrence of these cascade events can be considered as a fundamentally new physical phenomenon. As a confirmation of this behaviour, recent works by Moiser-Boss [14, 15] demonstrate the presence of high-energy charged particles emitted during experiments of Pd/D co-deposition. The employed solid state detector revealed the presence of alpha particles characterized by an energy higher than the usual one emitted during radioactive decay processes. Thus, it can be assumed that the excitation energy threshold, which the nuclear fragments or alpha particles must have to overcome the potential barrier, is sufficiently high to be detected by the experimental device used during the tests performed by the authors.

5.7 Conclusions

In previous papers by the authors, neutron emission measurements performed on Luserna stone specimens in compression were shown. From these experiments, it can be clearly seen that piezonuclear fission reactions giving rise to neutron emissions are possible in inert non-radioactive solids. In particular, during compression tests of specimens of sufficiently large size, the neutron flux was found to be of about one order of magnitude higher than the background level at the time of catastrophic failure. Further neutron emission measurements were carried out on Carrara marble specimens. Even if no significant neutron emissions were detected,

Table 5.4 Carrara Marble compression tests under monotonic displacement control. Average specimen dimensions and alpha emissions

Specimens	Dimension Diameter [mm]	Slenderness λ	Piston velocity [m/s]	Volume [mm^3]	Average peak load [kN]	Average alpha background (10^{-2} cps)	Maximum count rate at the alpha emission (10^{-2} cps)
A1	50	2	2×10^{-6}	196,250	154.9	12.00 ± 3.00	46.00 ± 12.00
A2	100	0.5	5×10^{-7}	392,500	810.8	12.00 ± 3.00	61.00 ± 15.00
A3	100	1	1×10^{-6}	785,000	737.9	12.00 ± 3.00	110.00 ± 28.00

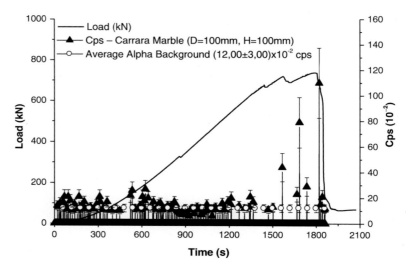

Fig. 5.5 Specimens A3. Load versus time diagrams, and alpha emission counting rate

important piezonuclear fission reactions were observed, also confirmed by considerable alpha particle emissions up to one order of magnitude higher than the background level. The compositional analysis by X-ray Photoelectron Spectroscopy (XPS) confirms the piezonuclear fission reactions involving the transmutation of calcium, magnesium, and oxygen into carbon. As in the case of Luserna stone, this hypothesis seems to find a surprising evidence and confirmation at the Earth crust scale. As a matter of fact, the piezonuclear reactions could interpret the most significant geophysical and geological transformations, today still unexplained.

References

1. Carpinteri A, Cardone F, Lacidogna G (2010) Energy emissions from failure phenomena: mechanical, electromagnetic, nuclear. Exp Mech 50:1235–1243
2. Carpinteri A, Borla O, Lacidogna G, Manuello A (2010) Neutron emissions in brittle rocks during compression tests: monotonic vs cyclic loading. Phys Mesomech 13:268–274
3. Carpinteri A, Lacidogna G, Manuello A, Borla O (2011) Energy emissions from brittle fracture: neutron measurements and geological evidences of piezonuclear reactions. Strenght Fract Complex 7:13–31
4. Albertini G, Calbucci V, Cardone F, Fattorini G, Magnani R, Petrucci A, Ridolfi F, Rotili A (2013) Evidence of alpha emission from compressed steel bars. Int J Mod Phys B 27 (23):1350124
5. Widom A, Swain J, Srivastava YN (2013) Neutron production from the fracture of piezoelectric rocks. J Phys G Nucl Part Phys 40:015006 (1–8)
6. Widom A, Swain J, Srivastava YN (2015) Photo-disintegration of the iron nucleus in fractured magnetite rocks with magnetostriction. Meccanica 50:1205–1216. doi:10.1007/s11012-014-0007-x

7. Bubble Technology Industries (1992) Instruction manual for the bubble detector. Bubble Technology Industries, Ontario
8. National Council on Radiation Protection and Measurements (1971) Protection against neutron radiation, NCRP Report 38
9. Cardone F, Cherubini G, Petrucci A (2009) Piezonuclear neutrons. Phys Lett A 373:862–866
10. Cardone F, Mignani R, Petrucci A (2009) Piezonuclear decay of thorium. Phys Lett A 373:1956–1958
11. Carpinteri A, Lacidogna G, Manuello A, Borla O (2013) Piezonuclear fission reactions from earthquakes and brittle rocks failure: evidence of neutron emission and non-radioactive product elements. Exp Mech 53:345–365
12. Ni M, Ratner BD (2008) Differentiating calcium carbonate polymorphs by surface analysis techniques – an XPS and TOF-SIMS study. Surf Interface Anal 40:1356–1361
13. Postendörfer J, Butterweck G, Reineking A (1994) Daily variation of the radon concentration indoors and outdoors and the influence of meteorological parameters. Health Phys 67:283–287
14. Mosier-Boss PA et al (2007) Use of CR-39 in Pd/D co-deposition experiments. Eur Phys J Appl Phys 40:293–303
15. Mosier-Boss PA et al (2010) Comparison of Pd/D co-deposition and DT neutron generated triple tracks observed in CR-39 detectors. Eur Phys J Appl Phys 51:20901–20911

Chapter 6
Elemental Content Variations in Crushed Mortar Specimens Measured by Instrumental Neutron Activation Analysis (INAA)

Alberto Carpinteri, Oscar Borla, and Giuseppe Lacidogna

Abstract Previous investigations concerning neutron emission measurements highlighted piezonuclear fission reactions during mechanical tests on iron-rich materials. Based on our experimental evidences, iron can be considered one of the most convenient elements as regards fission into aluminium or into magnesium and silicon. In the present investigation, we apply the Instrumental Neutron Activation Analysis (INAA) in order to provide experimental evidence of elemental content variations in mortar specimens subjected to compression tests up to crushing failure. To emphasize such a phenomenon, the specimens were highly enriched with iron oxides. Twenty-four chemical elements, including iron, aluminium, magnesium, and silicon, were quantified before and after the mechanical tests by means of chemical and INAA analyses. Our intention was mainly that of confirming low energy nuclear reactions involving fission of iron into aluminum. To this purpose, the concentrations of aluminum before and after the compression tests of the mortar specimens are presented and discussed.

Keywords Compression tests • Mortar • Piezonuclear fission reactions • Neutron activation analysis

6.1 Introduction

The results presented in [1–3] and related to compression tests on Luserna stone samples, have demonstrated that neutron emissions detected from specimens of sufficiently large dimensions exceeded the background level by approximately one order of magnitude at the moment of specimen failure.

A. Carpinteri (✉) • O. Borla • G. Lacidogna
Department of Structural, Geotechnical and Building Engineering,
Politecnico di Torino, Corso Duca degli Abruzzi 24, 10129 Torino, Italy
e-mail: alberto.carpinteri@polito.it

© Springer International Publishing Switzerland 2015
A. Carpinteri et al. (eds.), *Acoustic, Electromagnetic, Neutron Emissions from Fracture and Earthquakes*, DOI 10.1007/978-3-319-16955-2_6

These emissions are due to piezonuclear reactions involving fission of iron into alumium, or into magnesium and silicon. These assumed fissions are supported by spectroscopical analyses of the fracture surfaces. The results of Energy Dispersive X-ray Spectroscopy (EDS), performed on samples coming from the Luserna stone specimens used in the preliminary experiments [1–3], show that on the fracture surfaces a considerable reduction in the iron content (~25 %) is very consistently counterbalanced by an increment in Al, Si, and Mg concentrations [4]. A theoretical explanation about piezonuclear fission reactions has been provided very recently [5, 6].

In the present paper – after reporting an interesting experimental result on a single prismatic specimen ($4 \times 4 \times 16$ cm^3) of cementitious mortar enriched with iron – we describe new experimental evidences of piezonuclear reactions obtained on cubic mortar specimens (of 1.00 cm side) enriched with iron oxide and subjected to mechanical loading. The purpose is mainly that of analyzing the nuclear reaction, $Fe^{56} \rightarrow 2Al^{27} + 2n$, involving the symmetrical fission of iron into two atoms of aluminum during the compression test. The compression tests were carried out at the Fracture Mechanics Laboratory of the Politecnico di Torino. During the compression test of the single prismatic specimen ($4 \times 4 \times 16$ cm^3), a neutron emission approximately three times higher than the average natural background was observed at the moment of brittle fracture. On the other hand, due to the small dimensions of the cubic specimens, no relevant neutron emissions were detected during mechanical loading. In any event, important piezonuclear evidences were observed in the crushed mortar samples. They were analyzed by the Instrumental Neutron Activation Analysis (INAA) [7] measuring the γ radiation emitted during the subsequent radioactive decay of the produced nuclei. This analytical technique offers positive features such as high accuracy, precision, sensitivity, and the possibility of multielemental analysis.

Compared to the EDS technique, that is able to analyze only the fracture surfaces, the INAA technique allows a detailed volumetric analysis for the determination of elements even in minimal amounts. For this reason, the two compositional analysis methods are considered as complementary techniques. Moreover, we are dealing with a non-destructive procedure: no complex operations like dissolution, digestion or other chemical treatments of the samples are required, reducing the possibility of sample contamination. The irradiations with neutrons were carried out in the Laboratory for Applied Nuclear Energy (LENA) of the University of Pavia in a TRIGA MARK II reactor (General Atomics) [8]. The samples were irradiated for different times with different neutron fluencies depending on matrix type and on the elements to be determined. The experimental data collected during the tests highlight a significant increment in Al^{27}, up to 66 % of the initial concentration. Since the measurement method based on INAA was validated with the Certified Reference Materials (CRMs) (NIST

2709 San Joaquin Soil and NIST 2704 Buffalo River Sediment), the detected variations can be associated to the piezonuclear reactions occurring during the failure test.

6.2 Instrumental Neutron Activation Analysis

The Instrumental Neutron Activation Analysis (INAA) is a nuclear analytical technique useful for qualitative and quantitative determination of trace elements and macro-constituents in different matrices.

A sample is subjected to a neutron flux and radioactive nuclides are produced. During the radioactive decay, the nuclides emit gamma rays whose energies are characteristic for each nuclide. The intensity of these gamma rays compared to those emitted by a standard nuclide allows a quantitative measure of the concentrations of the various elements.

The n-gamma reaction is the fundamental one in neutron activation analysis. For example, consider the following reaction:

$$Al^{27}(n, \gamma) \rightarrow Al^{28} + \beta^- + \gamma, \tag{6.1}$$

where Al^{27} is a stable isotope of aluminium, whereas Al^{28} is a radioactive isotope (with a half-life $t\frac{1}{2}$ of 139 s). The gamma ray emitted during the decay of the Al^{28} nucleus has an energy of 1778.9 keV and this gamma ray is characteristic for this nuclide.

On the other hand, the activity of a particular radionuclide, at any time t during an irradiation, can be calculated from the following equation:

$$A_t = \sigma_{act} \varphi \, N(1 - e^{-\lambda t}), \tag{6.2}$$

where A_t is the activity in number of decays per unit time, σ_{act} is the activation cross-section, φ is the neutron flux (usually given in number of neutrons cm^{-2} s^{-1}), N is the number of parent atoms, λ is the decay constant (number of decays per unit time), and t is the irradiation time. After the sample has been activated and the resulting gamma ray energies and intensities have been determined using a solid-state detector (usually germanium), gamma rays passing through the detector generate free-electrons. The number of electrons (current) is related to the energy of the gamma rays.

Each radioactive nuclide is also decaying during the counting interval and corrections must be considered for this decay. The standard form of the radioactive decay correction is

$$A = A_o e^{-\lambda t}, \tag{6.3}$$

where A is the activity at any time t, A_o is the initial activity, λ is the decay constant and t is time.

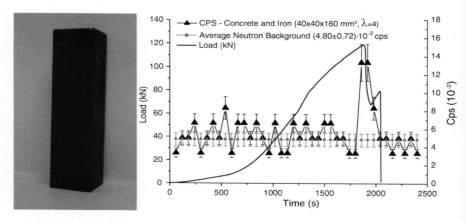

Fig. 6.1 Prismatic mortar specimen ($4 \times 4 \times 16$ cm^3) enriched with iron oxide (*left*). Load versus time diagrams, and neutron emission count rate (*right*)

6.3 Preliminary Test on a Single Prismatic Mortar Specimen

As anticipated in the Introduction, a prismatic cementitious mortar specimen enriched with iron, measuring $4 \times 4 \times 16$ cm^3 (Fig. 6.1, left), was tested under compression by an electronic controlled servo-hydraulic press with a maximum capacity of 1000 kN. This machine makes it possible to carry out tests in either load control or displacement control. The tests were performed by piston travel displacement control setting a velocity of 0.001 mm/s. Neutron emission measurements were made by means of a He^3 proportional counter [see Refs. 2, 3] placed at a distance of 10 cm from the specimen and enclosed in a polystyrene case, to prevent impacts and vibrations. The detector relies on a conversion process where an incident neutron interacts with a nucleus to produce a secondary charged particle. These charged particles are then detected, and from them the neutrons presence is deduced. In particular, the device was calibrated for the measurement of thermal neutrons; its sensitivity is 65 cps/$n_{thermal}$ (± 10 % declared by the factory), i.e., a thermal neutrons flux of 1 thermal neutron/s cm^2 corresponds to a count rate of 65 cps. The monitored neutron emissions (Fig. 6.1, right) exceeded the background value by approximately three times when the brittle failure occurred.

Although, the Neutron Activation Analysis considers only the iron isotope 58 while the investigated piezonuclear reaction regards Fe^{56}, important considerations can also be made for iron nuclei.

It is well-known that the natural abundance of Fe^{58} is 0.3 % (compared to 91.7 % of Fe^{56}). From the point of view of neutron activation, it is the isotope on which the concentration analysis of iron is based. Once subjected to a suitable neutron flux, the radioactive nuclide Fe^{59} is produced with a half life of about 45 days. This isotope is accompanied by three different gamma ray energy emissions, typical of this nuclide, and then iron concentration is calculated.

Fig. 6.2 Cubic mortar specimens (1.00 cm side) enriched with iron oxide (*left*). Mortar specimens during compression test. In front of the specimens the He^3 neutron detector was positioned at a distance of about 25 cm (*right*)

In particular, the INAA results for the fragments coming from the fracture surface show a lower concentration of iron (−14 %) if compared to the ones collected on the external surface of the specimen. This iron decrement appears to be counterbalanced by a rather close increment in aluminium (+19 %). Moreover, being the tested artificial material characterized by a sufficiently precise and homogeneous elemental distribution, the results of neutron activation analysis are even more accurate and reliable.

The results obtained in this preliminary test led us to consider specimens of smaller dimensions (1.00 cm^3) to verify, by means of the INAA (Instrumental Neutron Activation Analysis) technique, whether the neutron emissions were linked to compositional variations of the elements constituting the samples.

6.4 Tests on Cubic Mortar Specimens: Experimental Set-Up

Piezonuclear evidences were observed in seven of the cubic mortar specimens (of 1.00 cm side) enriched with iron oxide. In Fig. 6.2 (left) the specimens are shown. The tests were carried out by a MTS servo-hydraulic press, with a maximum capacity of 250 kN, working by a digital type electronic control unit. The force applied was determined by measuring the pressure in the loading cylinder by means of a transducer. The specimens were arranged (Fig. 6.2 right) with two surfaces in contact with the press platens, without coupling materials in-between, according to the testing modalities known as "test by means of rigid platens with friction". The platens were controlled with a wire-type potentiometric displacement transducer. Finally, the tests were performed under displacement control, with a planned displacement velocity equal to 0.001 mm/s.

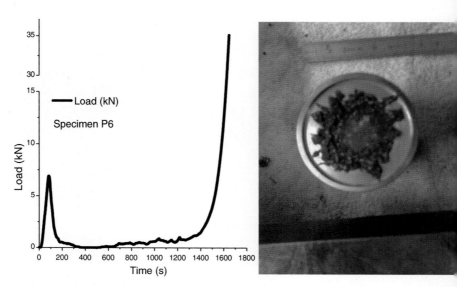

Fig. 6.3 Specimen P6, load versus time diagram (*left*). Specimen P6 after the crushing test (*right*)

During the experimental tests the full stress–strain curves were considered. The complete curve for specimen P6 is plotted in Fig. 6.3 (left). After the specimen has been crushed down to a heap of fragments (Fig. 6.3 right), the resistance to further deformation reaches a minimum after about 500 s from the beginning of the compression test. After that, the load begins to increase once more. The load increase continues until all the fragments are pulverized. The slope of the load versus displacement curve coincides with the testing machine stiffness. The test was stopped when a load five times higher than the peak load was reached.

In Table 6.1, experimental data concerning the compression tests on the seven mortar specimens are summarized.

6.5 Tests on Cubic Mortar Specimens: Experimental Results

In order to verify possible piezonuclear reactions, an analysis of the variations in the concentration of the isotope Al^{27} by means of INAA was carried out on the mortar specimens before and after the compression tests.

In Table 6.2, the chemical compositional analysis before the compression test on three of the seven tested specimens is reported.

Considering that – as explained in greater detail in [1–3] – we intend to detect the nuclear reaction, $Fe^{56} \rightarrow 2Al^{27} + 2n$, involving the symmetrical fission of iron into

Table 6.1 Compression tests under monotonic displacement control. Mortar specimen dimensions and neutron emissions

Specimens	Dimension Side [mm]	Piston velocity [m/s]	Average peak load [kN]	Average peak load at compaction [kN]	Average neutron background (10^{-2} cps)	Count rate at the neutron emission (10^{-2} cps)
P1	10	1×10^{-6}	2.69	33.45	3.73 ± 0.37	Background
P2	10	1×10^{-6}	7.11	52.66	3.73 ± 0.37	Background
P3	10	1×10^{-6}	6.50	44.94	3.73 ± 0.37	Background
P4	10	1×10^{-6}	7.55	30.23	3.73 ± 0.37	Background
P5	10	1×10^{-6}	7.89	34.66	3.73 ± 0.37	Background
P6	10	1×10^{-6}	6.89	35.27	3.73 ± 0.37	Background
P7	10	1×10^{-6}	5.71	31.14	3.73 ± 0.37	Background

Table 6.2 Mortar specimens chemical compositional analysis before the compression test

Chemical compositional analysis. Concentration in µg/g

Element	Sample 1	Sample 2	Sample 3	Mean value	Standard deviation %	Mass %
Mg	3,690	3,798	3,960	3,906	6.0	0.39
Ca	148,987	139,929	136,780	138,400	5.8	14
V	7.4	9.2	5.5	7.1	16	0.0007
Al	14,771	14,928	15,260	14,720	5.3	1.5
Mn	179	170	173	172	6.2	0.02
Si	244,616	270,000	255,663	254,424	3.4	25
Sm	0.72	0.62	0.66	0.67	7.1	6.7E-05
Mo	0.11	0.22	0.25	0.20	37	2.0E-05
U	0.51	0.43	0.65	0.53	21	5.3E-05
Br	0.25	0.21	0.21	0.23	9.3	2.3E-05
As	3.6	3.1	5.1	4.0	26	0.0004
K	6,601	6,689	5,032	6,107	15	0.61
La	5.0	4.7	4.4	4.7	6.0	0.0005
Ce	5.8	6.3	6.6	6.2	6.7	0.0006
Se	0.26	0.14	0.27	0.22	32	2.2E-05
Th	1.6	1.9	2.2	1.9	15	0.0002
Cr	32	30	34	32	6.5	0.003
Hf	0.74	0.88	0.96	0.86	13	8.6E-05
Cs	0.64	0.64	0.54	0.60	9.0	6.0E-05
Ni	12	12	12	12	0.34	0.001
Sc	0.68	0.69	0.92	0.76	18	7.6E-05
Rb	23	29	24	25	12	0.003
Fe	45,271	51,168	53,951	50,130	8.8	5.0
Ta	0.11	0.11	0.08	0.10	16	1.0E-05

two atoms of aluminum during the compression test, iron oxide was added to the samples to be analysed in order to emphasize such a nuclear reaction. An increase in the isotope Al^{27} was therefore expected.

In this framework, the seven cubic specimens of 1.00 cm side have been prepared. The elemental content of Al^{27} in each specimen was measured by INAA. After a cooling time of 6 months during which the induced radioactivity became negligible, the seven specimens were subjected to failure. After the failure test, the measurement of the elemental content of Al^{27} in each specimen was repeated.

The INAA feature of being a non-destructive technique allowed to measure the elemental content in the same specimen before and after the failure test. To quantify the Al^{27} content, the nuclear reaction Al^{27} (n, γ) $\rightarrow$ Al^{28} (t½ 139 s) was considered. The corresponding gamma emission occurs at 1778.9 keV. Samples, standards and CRMs were irradiated for 30 s with a neutron flux of $1 \times 10^{13} cm^{-2}$ s^{-1} in the pneumatic fast transfer thimble (Rabbit) of the reactor. The gamma spectrometry was carried out by counting the gamma emission for 300 s after a cooling time of 300 s. A home made standard solution was prepared as a comparator starting from a certified primary solution (Inorganic Ventures) of aluminium with a concentration of (9996 $\pm$ 56) µg/ml. The analytical method was checked by measuring the Al^{27} content of two Certified Reference Materials (NIST 2709 San Joaquin Soil and NIST 2704 Buffalo River Sediment).

The aluminium solution was prepared gravimetrically by pipetting aliquots of the certified primary solution onto a filter paper rolled up as a cylinder and inserted in polyethylene vials (Kartell). The same vials were also used for the mortar samples and the Certified Reference Materials (CRM). Samples, standards, CRMs and blanks (polyethylene vials) were inserted in containers for neutron irradiation.

The gamma counting facility consists of an HPGe detector coupled to a multichannel acquisition system (DSPEC from ORTEC-USA). The collected spectra were analyzed and processed with the Gammavision (ORTEC-USA) software package.

The results obtained during the INAA analyses are summarised in Table 6.3.

The experimental data collected highlight an average increase in the Al^{27} content within the samples. In particular, the results obtained show a significant increase in Al^{27} in three of the seven samples. The concentration of Al^{27} in Sample 1 is more than 50 % higher then the concentration of the same isotope before the failure test, whereas Samples 4 and 7 show an increase of 15 % and 8 %,

Table 6.3 Concentrations of Al^{27} and ratios before and after the failure test. The INNA relative concentration uncertainty due to the device sensitivity is to be considered equal to 0.04 %

	Concentration (in µg/g) of Al^{27} before and after the failure test						
	Sample 1	Sample 2	Sample 3	Sample 4	Sample 5	Sample 6	Sample 7
Before	14,490	14,500	14,880	13,230	15,000	15,090	13,440
After	24,110	14,590	15,570	15,160	14,880	14,630	14,380
After/Before	1.66	1.01	1.05	1.15	0.99	0.97	1.08

respectively. It is interesting to observe that no sample shows significant decrease in Al^{27} concentration. Since the measurement method based on INAA was validated with the CRMs, the detected variations can be associated to the assumed piezonuclear reactions occurring during the failure test.

6.6 Conclusions

Neutron emission measurements performed on Luserna stone specimens in compression were shown in previous papers by the authors [1–3]. From these experiments, it can be clearly seen that piezonuclear fission reactions giving rise to neutron emissions are possible in inert non-radioactive solids. In particular, during compression tests on specimens of sufficiently large size, the neutron flux was found to be of approximately one order of magnitude higher than the background level at the time of catastrophic failure.

In the framework of the present investigation, a preliminary compression test was conducted on a prismatic specimen ($4 \times 4 \times 16$ cm^3) realized by cementitious mortar enriched with iron. During the test a neutron emission of about three times higher than the average natural background was observed at the moment of final collapse. Based on this experiment, further neutron emission measurements were carried out on cubic mortar specimens (of 1.00 cm side) enriched with iron. In these samples, even if no significant neutron emissions were detected, important compositional changes were observed.

As a matter of fact, our intention was mainly that of analyzing the concentration of aluminum before and after the compression test to demonstrate the occurrence of low energy nuclear reactions involving the symmetrical fission of iron into two atoms of aluminum.

Differently from [3, 4], in the present investigation the compositional analysis of the cementitious mortar specimens was conducted with the INAA method that allowed elemental analyses within the entire volume of the samples, and not only on the fracture surfaces.

The Neutron Activation Analysis confirms the piezonuclear fission reaction $Fe^{56} \rightarrow 2Al^{27} + 2n$, involving transmutation of iron into aluminium, not only in natural rocks like Luserna stone, but also in artificial materials such as cementitious mortar.

Acknowledgements The Authors are grateful to Prof. S. L. Pagliolico from the Politecnico di Torino for the suggestions about the suitable composition of the cementitious mortar specimens enriched with iron oxide. The Authors express their gratitude also to Dr F. Canonico of Buzzi Unicem factory for taking care of manufacturing the mortar specimens.

They also wish to thank Dr L. Bergamaschi and Dr L. Giordani from the National Research Institute of Metrology (INRIM) for their assistance with the Instrumental Neutron Activation Analysis during the period 2009–2011.

References

1. Carpinteri A, Cardone F, Lacidogna G (2010) Energy emissions from failure phenomena: mechanical, electromagnetic, nuclear. Exp Mech 50:1235–1243
2. Carpinteri A, Borla O, Lacidogna G, Manuello A (2010) Neutron emissions in brittle rocks during compression tests: monotonic vs cyclic loading. Phys Mesomech 13:268–274
3. Carpinteri A, Lacidogna G, Manuello A, Borla O (2011) Energy emissions from brittle fracture: neutron measurements and geological evidences of piezonuclear reactions. Strenght Fract Complex 7:13–31
4. Carpinteri A, Chiodoni A, Manuello A, Sandrone R (2011) Compositional and microchemical evidence of piezonuclear fission reactions in rock specimens subjected to compression tests Strain 47(s2):282–292
5. Widom A, Swain J, Srivastava YN (2013) Neutron production from the fracture of piezoelectric rocks. J Phys G Nucl Part Phys 40:015006 (1–8)
6. Widom A, Swain J, Srivastava YN (2015) Photo-disintegration of the iron nucleus in fractured magnetite rocks with magnetostriction. Meccanica 50:1205–1216. doi:10.1007/s11012-014-0007-x
7. Alfassi ZB (1994) Chemical analysis by nuclear methods. Wiley, New York
8. The Laboratory of Applied Nuclear Energy ("LENA") (2013) http://www.unipv-lena.it/english-version/irradiation-service-a-analysis.html. Accessed June 2013

Chapter 7
Piezonuclear Evidences from Tensile and Compression Tests on Steel

Stefano Invernizzi, Oscar Borla, Giuseppe Lacidogna, and Alberto Carpinteri

Abstract Piezonuclear reactions concern neutron emissions triggered by high-frequency pressure-waves in inert, non-radioactive materials. This phenomenon has recently been detected in liquid solutions and brittle solids. In the present paper, the investigation is extended to a more ductile material, as steel is, subjected to different loading conditions. Although piezonuclear reactions are more likely to take place during the failure of brittle materials, the phenomenon is revealed also with the more ductile failure of metallic materials.

Keywords Piezonuclear reactions • Neutron emissions • Steel • Tensile and compression tests

7.1 Introduction

In recent years, neutron emission measurements were successfully performed on brittle solid specimens during crushing test failure. From those experiments, it can be clearly seen that piezonuclear reactions giving rise to neutron emissions are possible in inert, non-radioactive brittle materials like Luserna stone or concrete [1–5]. A theoretical explanation has been also provided in recent works [6, 7].

The term "piezonuclear" reaction refers to low energy nuclear reactions, which are triggered by the fundamental action of pressure. The phenomenon of neutron emission was firstly recognized in stable iron nuclides contained in aqueous solutions of iron chloride or nitrate subjected to cavitation induced by ultrasounds [8]. It has recently been detected also in solids during mechanical tests on laboratory specimens. On the other hand, quantitative Scanning Electron Microscope (SEM) analyses of the fracture surfaces of broken rock specimens showed that different kinds of element transmutation could be the results of fission reactions generating neutron emission [9, 10].

S. Invernizzi (✉) • O. Borla • G. Lacidogna • A. Carpinteri
Department of Structural, Geotechnical and Building Engineering, Politecnico di Torino, Corso Duca degli Abruzzi 24, 10129 Torino, Italy
e-mail: stefano.invernizzi@polito.it

In the present paper, the investigation is extended to a more ductile metallic material, like steel, subjected to different testing conditions.

Steel bars subjected to uniform tensile or compressive loading, as well as notched specimens subjected to the Charpy test are described. From these tests positive evidences of neutron emission are obtained, which depend on the loading condition, the specimen size, and, therefore, the brittleness of the mechanical response [11].

For an accurate neutron evaluation, a He^3 proportional counter (Xeram, France) and a He^3 radiation monitor AT1117M (ATOMTEX, Minsk, Republic of Belarus) were used. Since neutrons are electrically neutral particles, they cannot directly produce ionization in a detector, and therefore cannot be directly detected. This means that neutron detectors must rely upon a conversion process where an incident neutron interacts with a nucleus to produce a secondary charged particle. These charged particles are then detected, and from them the neutrons presence is deduced. In addition, a set of passive neutron detectors (BDT Bubble Dosimeter Thermal and BD-PND Bubble Dosimeter-Personal Neutron Dosimeter models) insensitive to electro-magnetic noise and with zero gamma sensitivity was used. The dosimeters, based on superheated bubble detectors (BTI, Ontario, Canada) (Bubble Technology Industries (1992)) [12], were calibrated at the factory against an AmBe source in terms of NCRP38 (NCRP report 38 (1971)) [13].

Semi-quantitative results from XRF (X-Ray Fluorescence spectroscopy) and SEM analyses are also described. Due to the size of the steel bars, a portable XRF analysis has been carried out to characterize the chemical composition of the fracture surface of the broken specimens. On the other hand, since the Charpy test specimens are smaller enough, they have been analyzed by SEM.

Although piezonuclear reactions are more likely to take place during the failure of brittle materials, the phenomenon has been revealed also with the more ductile failure of metallic materials.

7.2   Steel Specimens Under Tensile and Compression Loading

Specific tests were conducted on 18 steel specimens of different diameter subjected to tensile stress, and on two specimens subjected to compression stress condition [11].

Regarding the 18 specimens tested under tensile condition, 9 had variable diameter ranging from 14 to 28 mm, whereas the remaining nine had the same diameter equal to 40 mm and a length of 1000 mm.

In Table 7.1, the applied loads, as well as the geometrical and mechanical characteristics for each of the first nine specimens are summarized. The specimens were subjected to tensile loading up to failure according to EN ISO 6892 recommendation [14].

Table 7.1 Tensile loading test on steel specimens

Tensile loading test

Bar type	Section S_0 (mm^2)	Yield strength limit P_y (kN)	Ultimate strength P_u (kN)	Yielding stress σ_y (MPa)	Stress peak σ_u (MPa)	Elongation at P_u, ε_u (%)	Elongation at failure ε_f (%)
D14_1	153.94	84.68	96.81	550.11	628.88	25.56	37.71
D20_1	314.16	160.36	194.17	510.45	618.06	34.39	53.72
D20_2	314.16	166.61	193.24	530.33	615.10	39.28	54.81
D22_1	380.13	148.52	182.35	390.71	479.70	27.71	46.07
D22_2	380.13	155.39	188.08	408.78	494.78	33.72	52.26
D24_1	452.39	230.00	306.41	508.66	677.65	38.18	58.99
D24_2	452.39	290.05	337.10	641.15	745.15	31.36	50.74
D28_1	615.44	351.42	437.58	571.02	711.00	37.96	61.27
D28_2	615.44	362.30	441.62	588.69	717.57	36.69	55.77

Table 7.2 Tensile loading test on steel specimens with a diameter of 40 mm

Tensile loading test

Bar type	Section S_0 (mm^2)	Ultimate strength P_u (kN)	Stress peak σ_u (MPa)
D40	1256	797.0	634.55
D40	1256	799.8	636.78
D40	1256	800.7	637.50
D40	1256	791.6	630.25
D40	1256	800.7	637.50
D40	1256	801.9	638.46
D40	1256	811.9	646.42
D40	1256	837.0	666.40
D40	1256	838.7	667.75

To carry out these experiments, an hydraulic press, Walter Bai type, with electronic control was used. The test was conducted in three subsequent stages. In the first stage it was controlled by stress increments of 15 MPa/s up to the value of about 500 MPa, which corresponds to the yield stress of the material. Subsequently, the test was controlled by an imposed strain of 0.16 mm/s up to an elongation equal to 10 % of the initial length. In the last stage, which ended with the specimen failure [11], the imposed deformation was applied by displacement increments of 0.33 mm/s.

The additional tests conducted on the nine steel specimens with constant diameter and length (40 and 1000 mm, respectively) were carried out in the framework of the usual procedures, according to which only the ultimate strength was registered.

In Table 7.2, the applied loads, as well as the specimen geometrical characteristics are summarized. An average loading ramp of about 323 kN/min was applied and a mean ultimate strength of 808.81 kN was observed.

Fig. 7.1 (a) Experimental set-up for the compression test performed on specimen C1 using the Galdabini hydraulic press. (b) Specimen C1 at the end of the test [11] (observe the presence of cracks on the lateral surface of the specimen)

Table 7.3 Compressive loading test on steel specimens

Compression Test	
Specimen C1	
Section S_0	1256 (mm^2)
Yield strength limit, P_y	638.56 (kN)
Ultimate strength, P_u	850.31 (kN)
Specimen C2	
Section S_0	1256 (mm^2)
Yield strength limit, P_y	607.21 (kN)
Ultimate strength, P_u	808.69 (kN)

To carry out the tensile test on the specimens with a diameter of 40 mm, an hydraulic press (METRO COM), with manual control and a maximum load of 1000 kN was used. Each test was conducted in a single stage up to the specimen failure.

Concerning the trials in compression loading condition, two specimens with diameter of 40 mm were tested using an hydraulic press, Galdabini type, with a maximum load of 5000 kN (Fig. 7.1a). A loading ramp of 58 kN/min was applied and the tests stopped after about 30 min, at a load of 2000 kN, corresponding to a specimen shortening approximately of 50 % (see Fig. 7.1b). Considering a symmetric behaviour of steel in tensile and compression conditions, the yield strength (638.56 kN) limit and the ultimate strength (850.31 kN) of specimen C1 were identified considering the yielding and the ultimate strength obtained during the tensile loading test. Similar results were also obtained for specimen C2 (see Table 7.3).

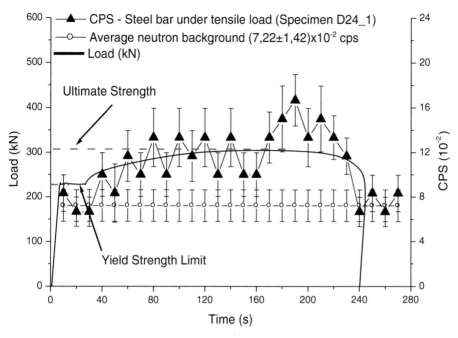

Fig. 7.2 Load versus time diagrams and neutron emission measurements for specimen D24_1

7.3   Neutron Emission Detection

As a first hint, the measurement of natural neutron background, for each test described in the following, was performed in order to assess the average background affecting data acquisition at experimental room conditions. He^3 devices (thermal neutron counter and ATOMTEX) and thermodynamic bubble detectors (BDT and BD-PND types) were employed to monitor the typical daily fluctuation of background level. Additional background measurements, for a period of at least 1 h, were repeated before each test in order to check a possible variation in natural background. Moreover, to provide a precise neutron evaluation in terms of CPS (counts per second), a suitable reference time was chosen depending on the experimental condition. It was fixed up to 60 s to reduce the probability to detect zero counts during neutron emission evaluation.

The tensile tests on the nine steel bars reported in Table 7.1 and the compression tests performed on specimens C1 and C2 were monitored by the He^3 (Xeram, France) neutron detector. The He^3 proportional counter was placed close to the experimental set-up in both tests. In Fig. 7.2, the load vs. time diagram for specimen D24_1 is reported with the neutron emission measurements. The average neutron background level measured before the test was equal to $(7.22 \pm 1.42) \times 10^{-2}$ cps. It can be noted that during the test, and in particular in correspondence to the achievement of the yield strength limit equal to 230 kN (see Fig. 7.2), neutron emissions increased up to $(11.67 \pm 2.29) \times 10^{-2}$ cps. The increment was about 1.5 times higher than the background level. In addition, in correspondence to the ultimate strength (306 kN) a maximum neutron emission of $(16.67 \pm 2.29) \times 10^{-2}$ cps was measured. This last emission level corresponds to an increment more than twice the background level measured before the experiment. Finally, after the steel bar failure, the neutron emissions decreased almost instantaneously down to the background level measured before the experiment.

As regards the tensile loading tests on the other steel bars reported in Table 7.1, a similar behaviour in neutron emission measurements was evidenced for specimens characterized by diameters equal to or greater than 24 mm. In the following (see Fig. 7.3) the load versus time diagrams are reported together with the neutron emission measurements also for the specimens D20, D22, and D28. For the largest specimens (D28), neutron emissions reached values up to three times higher than the average natural background.

The remaining nine tensile loading tests on specimens with a diameter of 40 mm have been monitored by the ATOMTEX He^3 neutron device and by 6 bubble neutron detectors, 3 BDT and 3 BD-PND. The average neutron background level measured by bubble detectors before the test was equal to (0.063 ± 0.016) μSv/h. The bubble dosimeters were placed all around the specimens while the He^3 detector was positioned at a distance of about 50 cm from the steel bar. A neutron emission up to (0.115 ± 0.029) μSv/h, corresponding to an increase more than twice the background level, was monitored at the end of the test only in the low energy range. This behaviour is also confirmed by the data acquired by the Atomtex device. In fact no significant increase in the high energy neutron field was observed.

For specimen C1, the load versus time diagram is reported in Fig. 7.4. Also in this case, the neutron emissions show an appreciable increase immediately after the achievement of the ultimate strength (850 kN). At this point, the maximum neutron emission level $(19.99 \pm 2.96) \times 10^{-2}$ cps corresponds to an increment of about three times the background level. As can be seen from Fig. 7.1b, the final section area of specimen C1 is sensibly larger than the initial nominal area, so that the theoretical ultimate strength was widely overcome.

For sample C1, similarly to specimen D24_1, it is also possible to observe that the maximum neutron emission level, reached after the ultimate strength and equal to $(19.10 \pm 2.29) \times 10^{-2}$ cps, corresponds to an increment of about three times with respect to the same background level [11].

It is important to emphasize that the emission of neutrons is a burst-like phenomenon due to three main reasons. The first is that local propagation (or redistribution) of defects is discontinuous and takes place similarly to the stick-slip phenomenon in friction. The second is due to the directionality of the neutron emission, that influences the probability of being acquired by the instrumentation. Finally, it depends on the experimental condition and on the reference time window over which the neutron counting rate is measured. If the reference time is not sufficiently long, the probability exists to detect zero counts in the chosen time window. For this reason, if the experimental conditions allow it (i.e. the duration of the test), it is better to set a reference time window sufficiently long to monitor an appreciable quantity of events.

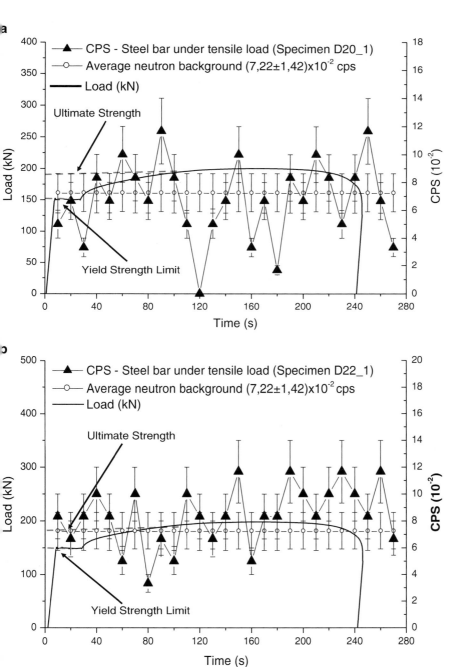

Fig. 7.3 Load versus time diagrams and neutron emission measurements for specimens D20 (**a**), D22 (**b**), and D28 (**c**) under tensile loading tests

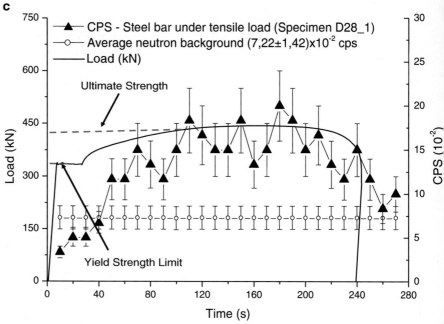

Fig. 7.3 (continued)

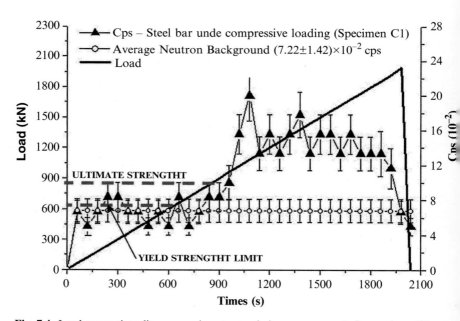

Fig. 7.4 Load versus time diagrams and neutron emission measurements for specimen C1

7.4 XRF Analysis of Steel Rebars

The emission of neutrons is usually due to nuclear phenomena, such as fission and spallation, which yield products different from the atomic elements that were present prior to the test. Those elements can be both radioactive or not, like in the case of the so-called cold fission or low energy fission processes [15].

Due to the size of the steel rebars, one of the most affordable techniques is portable XRF (X-Ray Fluorescence), which is commonly adopted for the characterization of corrosion resistant steel rebars [16]. Unfortunately, the XRF technique is not able to detect light elements, with atomic number lower than 15.

XRF analyses were performed with an Assing LITHOS 3000 portable spectrometer equipped with a molybdenum tube and a Peltier cooled Si-PIN detector with a sensitive area of about 7 mm^2, and a berillium (Be) window 12.5 μm thick. For the measurements on the Graduale the X-ray tube voltage was 25 kV, the current was 50 μA, and the acquisition time was 720 s. A 10 mm collimator was used for an investigated area of approximately 5 mm in diameter.

The acquisition was carried out adopting a calibration profile for metals, both on the cutting surface and on the fracture surface. Several acquisition points were selected, close to the bar axis as well as close to the lateral surface. Finally, acquisitions on the lateral surface were also performed.

Figure 7.5 shows the details of the location of the acquisition points.

Figure 7.6 shows the scattered X-ray spectrum, for the different acquisition points, that allows for the element tracing. As expected, iron (Fe) is by a great extent the most abundant element.

The spectral analysis allows for the determination of additional elements, such as Mn, Cr, Mo, Ni, and Cu, which are frequent impurities in common rebar steel. In addition, a quite well detectable peak in the spectrum is present in correspondence to Scandium (Sc) with an approximate concentration of 50 ÷ 100 ppm. Scandium is a quite rare element on the Earth's crust; therefore the presence of the peak is curious. Moreover, the magnitude of the peak seems greater on the fracture surface, especially closer to the lateral surface, rather than on the cutting surface, as shown in the comparison spectrum of Fig. 7.6c. This suggests that the presence of scandium could be somehow linked to the level of strain experienced by the material.

It is worth noting that the adopted portable XRF analyzer is not suitable to detect light elements as aluminum (Al), which is expected as piezonuclear fission product from iron. Nevertheless, scandium is detected from photofission experiments on iron [17], as well as in iron meteorites due to the effect of exposition to cosmic radiation [18]. Therefore, although further analyses are necessary, the presence of scandium could be a clue of the presence of other lighter fission products. For example, the piezonuclear transmutation of Fe into Sc should be accompanied by the presence of Boron (B).

Fig. 7.5 XRF acquisition points on the steel rebar: on the cutting section (**a**), on the fracture surface (**b**), on the lateral surface (**c**)

The acquisition on the lateral surface (points B6 and B7) did not reveal any anomalous element, a part from a much higher presence of Ni and Cu, which is likely due to surface passivation treatments.

7.5 Charpy Test

The Charpy test was carried out according to the European code EN 10045/1-1990 [19]. The nominal energy of the adopted Charpy pendulum was 300 J. Five standard steel specimens with a V-notch and a (10×8) mm^2 section were analyzed. The second specimen was preventively refrigerated in liquid nitrogen to increase its fragility. In this case, the dissipated energy during the test was about 6 J, and the specimen split into two separated halves.

The other specimens, characterized by different carbon content and obtained from slightly different thermal treatment, dissipated energies ranging from 76 to 175 J, and did not separate completely after the pendulum impact.

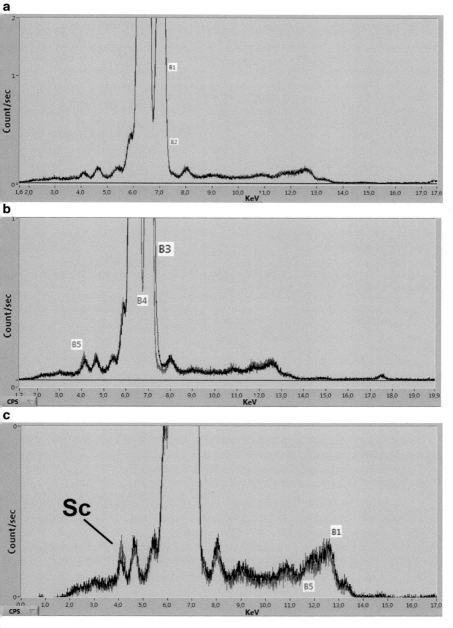

Fig. 7.6 XRF spectra: acquisition on the cutting section points B1 and B2 (**a**); acquisition on the fracture surface points B3, B4 and B5 (**b**); comparison between points B1 and B5 (**c**)

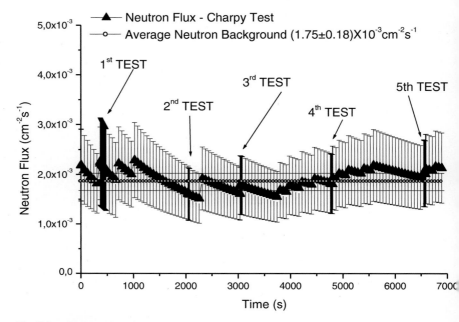

Fig. 7.7 Diagram of the Atomtex neutron acquisition during the Charpy tests

The Charpy tests were performed with approximately 30 min interval between each other, in order to allow for stable bubble nucleation in the bubble detector sensors. The analysis performed did not reveal any sensible bubble nucleation in any of the sensors, after each test. Contemporarily, the He^3 Atomtex acquisition was carried out.

Figure 7.7 shows that, in correspondence to the third and the fifth test, the recorded neutron emission was well above the background mean level. Nevertheless, such difference is still comprised in the confidence bar.

In particular, it appears that the greater variations from the background level were recorded in correspondence to the first (150 J dissipated), the third (175 J dissipated), and the fifth test (135 J dissipated). This suggests that brittleness is not the only governing parameter. In fact, in order to be able to detect neutron emissions with the adopted instrumentations, a certain amount of energy dissipation has to take place. This is not the case of the very brittle refrigerated sample two, which provided only 6 J, with no detectable neutron emissions.

In order to obtain evidences of possible transmutations of iron, consequence of piezonuclear reactions, the specimens were analyzed by SEM (EDAX), at the Department of Material Science of the Politecnico di Torino. The instrument can perform the Energy Dispersive X-Ray Spectroscopy. The volume of material involved in the measurement in the present case can be estimated equal to 1 cubic millimeter, based on the energy of the laser beam and on the size of the acquisition spot (30×30) μm^2.

Table 7.4 Quantitative spectrography for the third specimen

Location	Weigth %				
	Fe	Si	Mn	Al	Ca
Fracture surface	98.50	0.39	1.11	–	–
	98.73	0.29	0.99	–	–
	98.57	0.24	1.19	–	–
	98.62	0.36	1.02	–	–
	98.42	0.41	1.17	–	–
	98.09	0.35	1.57	–	–
V notch surface	98.37	0.47	1.16	–	–
	98.23	0.47	1.30	–	–
Impact area	98.36	0.63	1.02	–	–
	97.53	0.96	1.52	–	–
	97.59	1.32	1.10	–	–
Inclusion	70.60	11.07	–	10.27	8.06

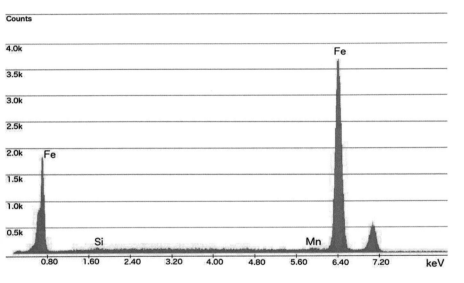

Fig. 7.8 Typical isotopes spectrum, where no anomalous elements are detected

Different areas of each specimen were analyzed, respectively in correspondence to the crack surface near the V notch (brittle propagation), or quite far from it (more stable propagation). In addition, the lateral surfaces of the specimens, and the area in correspondence to the impact were investigated. Most of the analyzed surfaces were rough, thus the acquisition was not in the ideal instrumental condition.

The quantitative element composition measured for the third specimen is reported in Table 7.4. An example of the element spectrum is shown in Fig. 7.8. No clear evidence of anomalous elements was recorded in correspondence to the

a

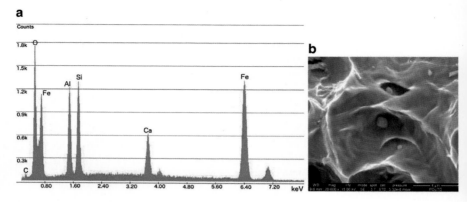

Fig. 7.9 Acquisition in correspondence to an inclusion: elements spectrum (**a**), micrography (**b**)

fracture surface, to the lateral surface, or to the impact region. On the other hand, Fig. 7.9a shows the spectrum when the acquisition is performed in the vicinity of an inclusion, corresponding to the acquisition area shown in Fig. 7.9b. In this case, the percentage composition is different, but this could be due to the original composition of the inclusion.

A small neutron emission was recorded, which was not completely distinguishable from the background. The metallic crystalline structure does not appear to be particularly prone to the piezonuclear phenomena. In addition, the small volumes and dissipated energies, combined with some difficulties in the positioning of the sensors, negatively affect the chance of detecting neutron emissions.

7.6 Conclusions

In the present paper, an investigation about piezonuclear neutron emissions from ductile materials, like steel, subjected to different loading conditions, is presented. Steel bars subjected to uniform tensile loading, as well as notched specimens subjected to the Charpy test, and uniform compression tests were considered. Some positive evidences of neutron emission are reported, which depend on the loading condition, the size and the brittleness of the specimen, and also on the amount of the dissipated mechanical energy. The XRF analysis of the fracture surface of steel bars broken in tension revealed the presence of an anomalous element, i.e. Scandium, which is very rare on the Earth's crust, whereas it is often found on iron subjected to photodisintegration (artificial irradiation or cosmic radiation in meteorites). Further investigations are necessary in order to combine portable XRF with other techniques (e.g. inductively coupled plasma mass spectrometry) able to detect lighter elements, which are expected as products of piezonuclear fissions.

Acknowledgements We want to deeply acknowledge the help of Prof. JeanMarc Tulliani (Politecnico di Torino) for the SEM analysis and Charpy tests, as well as the collaboration of Dr. Marco Nicola (Adamantio srl, c/o Incubatore d'Impresa dell'Università di Torino, via Quarello 11/a, Torino, Italy) for the semi-quantitative XRF analyses of steel rebars.

References

1. Carpinteri A, Cardone F, Lacidogna G (2009) Piezonuclear neutrons from brittle fracture: early results of mechanical compression tests. Strain 45:332–339
2. Cardone F, Carpinteri A, Lacidogna G (2009) Piezonuclear neutrons from fracturing of inert solids. Phys Lett A 373:4158–4163
3. Carpinteri A, Cardone F, Lacidogna G (2010) Energy emissions from failure phenomena: mechanical, electromagnetic, nuclear. Exp Mech 50:1235–1243
4. Carpinteri A, Borla O, Lacidogna G, Manuello A (2010) Neutron emissions in brittle rocks during compression tests: monotonic vs cyclic loading. Phys Mesomech 13:268–274
5. Carpinteri A, Lacidogna G, Manuello A, Borla O (2011) Energy emissions from brittle fracture: neutron measurements and geological evidences of piezonuclear reactions. Strength Fract Complex 7:13–31
6. Widom A, Swain J, Srivastava YN (2013) Neutron production from the fracture of piezoelectric rocks. J Phys G Nucl Part Phys 40:015006 (8 pp)
7. Widom A, Swain J, Srivastava YN (2015) Photo-disintegration of the iron nucleus in fractured magnetite rocks with magnetostriction. Meccanica 50:1205–1216
8. Cardone F, Cherubini G, Petrucci A (2009) Piezonuclear neutrons. Phys Lett A 373:862–866
9. Carpinteri A, Chiodoni A, Manuello A, Sandrone R (2011) Compositional and microchemical evidence of piezonuclear fission reactions in rock specimens subjected to compression tests. Strain 47(s2):282–292
10. Carpinteri A, Manuello A (2010) Geomechanical and geochemical evidence of piezonuclear fission reactions in the Earth's crust. Strain 47(2):267–281
11. Carpinteri A, Lacidogna G, Manuello A, Borla O (2013) The phenomenon of neutron emission from earthquakes and brittle rocks failure: mechanical tests, microchemical analyses, geological consequences. Exp Mech 53:345–365
12. Bubble Technology Industries (1992) Instruction manual for the Bubble detector, Chalk River
13. National Council on Radiation Protection and Measurements (1971) Protection against neutron radiation, NCRP report 38
14. EN ISO 6892–1:2009 (2009) Metallic materials – tensile testing – part 1: method of test at room temperature
15. Gönnenwein F, Börsig B (1991) Tip model of cold fission. Nucl Phys A 530(1):27–57
16. Sharp SR, Lundy LJ, Nair H, Moen CD, Johnson JB, Sarver BE (2011) Acceptance procedures for new and quality control procedures for existing types of corrosion resistant reinforcing steel. Virginia Department of Transportation Innovation and Research, Virginia
17. Fulmer CB, Toth KS, Williams IR, Handley TH, Dell GF, Callis EL, Jenkin TM, Wyckoff JM (1970) Photonuclear reactions in iron and aluminum bombarded with high-energy electrons. Phys Rev C 2(4):1371–1378
18. Wanke H (1960) Scandium-45 as reaction product of cosmic radiation in iron meteorites. II", Zeitschrift fuer Naturforschung (West Germany) Divided into Z. Nautrforsch., A, and Z. Naturforsch., B: Anorg Chem, Org Chem, Biochem Biophys 15a:953–964
19. EN 10045/1 Metallic materials – Charpy impact test Test method (1990) European Committee for Standardization. Rue de Stassart 36, Bruxelles

Chapter 8
Cold Nuclear Fusion Explained by Hydrogen Embrittlement and Piezonuclear Fissions in Metallic Electrodes: Part I: Ni-Fe and Co-Cr Electrodes

Alberto Carpinteri, Oscar Borla, Alessandro Goi, Amedeo Manuello, and Diego Veneziano

Abstract Several evidences of anomalous nuclear reactions occurring in condensed matter have been observed in the phenomenon of electrolysis. Despite the great amount of experimental results coming from the so-called Cold Nuclear Fusion research activities, the comprehension of these phenomena still remains unsatisfactory. On the other hand, as reported by most of the articles devoted to Cold Nuclear Fusion, one of the principal features is the appearance of microcracks on the electrode surfaces after the experiments. In the present paper, a mechanical explanation is proposed considering a new kind of nuclear reactions, the piezonuclear fissions, which are a consequence of hydrogen embrittlement of the electrodes during electrolysis. The experimental activity was conducted using a Ni-Fe anode and a Co-Cr cathode immersed in a potassium carbonate solution. Emissions of neutrons and alpha particles were measured during the experiments and the electrode compositions were analyzed both before and after the electrolysis, revealing the effects of piezonuclear fissions occurring in the host lattices. The symmetrical fission of Ni appears to be the most evident observation. Such reaction would produce two Si atoms or two Mg atoms with alpha particles and neutrons as additional fragments.

Keywords Cold nuclear fusion • Piezonuclear fissions • Hydrogen embrittlement • Electrolysis • Nickel

A. Carpinteri (✉) • O. Borla • A. Manuello • D. Veneziano
Department of Structural, Geotechnical and Building Engineering, Politecnico di Torino, Corso Duca degli Abruzzi 24, 10129 Torino, Italy
e-mail: alberto.carpinteri@polito.it

A. Goi
Private experimentalist Varese, Italy

© Springer International Publishing Switzerland 2015   99
A. Carpinteri et al. (eds.), *Acoustic, Electromagnetic, Neutron Emissions from Fracture and Earthquakes*, DOI 10.1007/978-3-319-16955-2_8

8.1   Introduction

During the last two decades several evidences of anomalous nuclear reactions occurring in condensed matter have been observed [1–34]. These tests were characterized by significant neutron and alpha particle emissions as well as by extra heat generation. At the same time, appreciable variations in the chemical composition after brittle or during fatigue fracture were detected [35–44].

Most relevant papers on the so-called Cold Nuclear Fusion describe broad experimental activities conducted on electrolytic cells powered by direct current and filled with ordinary or heavy water solutions. In particular, in 1989, Fleischmann and Pons proposed the first experiment reproducing Cold Nuclear Fusion by means of electrolysis [6]. They asserted that the Palladium electrode reacted with the deuterium coming from the heavy water solution [6]. Later works reported that Pt and Ti electrodes had also been electrolyzed with D_2O to produce extra energy and chemical elements previously absent [26, 28, 29]. Extra energy has been also produced from electrolysis with Ni cathodes and H_2O-based electrolyte [13]. Furthermore, it was affirmed that a voltage sufficient to induce plasma generates a large variety of anomalous nuclear reactions when Pd, W, or C cathodes are adopted [16, 21–25].

In many of these experiments, the generated heat was calculated to be several times the input energy and the neutron emissions rate, during electrolysis, was measured to be about three times the natural background level [6]. In 1998, Mizuno presented the results of the measurements conducted by means of neutron emission detectors and compositional analysis techniques related to different electrolytic experiments [22]. A relevant heat generation was observed when the cell was supplied with high voltage, with an excess energy of 2.6 times the input one. Remarkable neutron emissions were revealed during these tests, as well as a considerable amount of new elements, i.e. Pb, Fe, Ni, Cr, and C, with the isotopic distribution of Pb deviating greatly from the natural isotopic abundances [22]. These results suggested that nuclear reactions took place during the electrolysis process [22]. Later, in 2002 Kanarev and Mizuno reported the results obtained from the surface compositional analysis of iron electrodes (99.90 % of Fe) immersed in KOH and NaOH solutions [34]. After the experiments, EDX spectroscopy revealed the appearance of several chemical elements previously absent. Concentrations of Si, K, Cr, and Cu were found on the surfaces of the operating cathode immersed in KOH. Analogously, concentrations of Al, Cl, and Ca were noticed on the iron electrode surfaces operating in NaOH. These findings are an evidence of compositional changes occurring during plasma formation in electrolysis of water [34]. In 2007, Mosier-Boss et al. [31, 33] obtained important proofs of anomalous measurements in experiments conducted by electrolytic co-deposition cells. More in detail, anomalous effects observed in the Pd/D system include heat and helium-4 generation, tritium, neutrons, gamma/X-ray emissions, and transmutations [7, 12, 31, 33].

As written by Preparata, "despite the great amount of experimental results observed by a large number of scientists, a unified interpretation and theory of these phenomena has not been accepted and their comprehension still remains unsolved" [6–9, 26, 27]. On the other hand, as shown by most articles devoted to Cold Nuclear Fusion, one of the principal features is the appearance of microcracks on electrode surfaces after the tests [26, 27]. Such evidence might be directly correlated to hydrogen embrittlement of the material composing the metal electrodes (Pd, Ni, Fe, Ti, etc.). This phenomenon, well-known in Metallurgy and Fracture Mechanics, characterizes metals during forming or finishing operations [45]. In the present study, the host metal matrix (for example Ni or Pd) is subjected to mechanical damaging and fracturing due to external atoms (deuterium or hydrogen) penetrating into the lattice structure and forcing it, during the gas loading. Hydrogen effects are largely studied especially in metal alloys, where the presence of H free atoms in the host lattice causes the metal to become more brittle and less resistant to crack formation and propagation. In particular, hydrogen generates an internal stress that lowers the fracture stress of the metal so that brittle crack growth can occur under a hydrogen partial pressure below 1 atm [45, 46].

Some experimental evidence shows that neutron emissions may be strictly correlated to fracture of non-radioactive or inert materials. From this point of view, anomalous nuclear emissions and heat generation had been verified during fracture in fissile materials [2–4] and in deuterated solids [5, 8, 30]. The experiments recently proposed by Carpinteri et al. and by Cardone et al. [35–43] represent the first evidence of neutron emissions due to piezonuclear fissions recognized during failure of inert, stable, and non-radioactive solids under compression, as well as from non-radioactive liquids under ultrasound cavitation [35, 36]. In particular damaged cratered zones have been observed in recent experiments reported by Albertini et al. where iron bars have been subjected to pressure waves. In this case the presence of elements such as O, Cl, K, Cu cannot be attributed to the occurrence of non-metallic inclusions or to contamination occurred during fabrications [42].

In the present paper, we analyze neutron and alpha particle emissions during tests conducted on an electrolytic cell, where the electrolysis is obtained using Ni-Fe and Co-Cr electrodes in aqueous potassium carbonate solution. Voltage, current intensity, solution conductivity, temperature, alpha and neutron emissions were monitored. The compositions of the electrodes were analyzed both before and after the tests. Strong evidences suggest that the so-called Cold Nuclear Fusion, interpreted under the light of hydrogen embrittlement, may be explained by piezonuclear fission reactions occurring in the host metal, instead of by the nuclear fusion of H isotopes adsorbed in the lattice. These new kind of fission reactions have recently been observed from the laboratory to the Earth's crust scale, when particular stress waves originate from fracture or fatigue phenomena, like in correspondence to an impending earthquake [38–40, 44].

8.2 Experimental Set-Up and Measurement Equipments

8.2.1 The Electrolytic Cell and the Power Circuit

Over the last 10 years, specific experiments have been conducted on an electrolytic reactor (owners: Mr. A. Goi et al.). The aim was to investigate whether the anomalous heat generation may be correlated to a new type of nuclear reactions during electrolysis phenomena. The reactor was built in order to be appropriately filled with a salt solution of water and Potassium Carbonate (K_2CO_3). The electrolytic phenomenon was obtained using two metal electrodes immersed in the aqueous solution. The solution container, named also reaction chamber in the following, is a cylinder-shaped element of 100 mm diameter, 150 mm high and 5 mm thick (Fig. 8.1a). For the reaction chamber, two different materials were used during the experiments: Pyrex glass and Inox AISI 316 L steel. The two metallic electrodes were connected to a source of direct current: Ni-Fe based electrode as positive pole (anode), and Co-Cr based electrode as negative pole (cathode) (see Fig. 8.1b).

With regard to the experiment described in the present paper, after approximately ten operating hours, the generation of cracks was observed on the glass container, which forced the authors to adopt a more resistant reaction chamber made of steel. Teflon lids are sealed to both the upper and the lower openings of the chamber. The reaction chamber base consists of a ceramic plate preventing the direct contact between liquid solution and Teflon lid (see Fig. 8.1a). Two threaded holes host the electrodes, which are screwed to the bottom of the chamber.

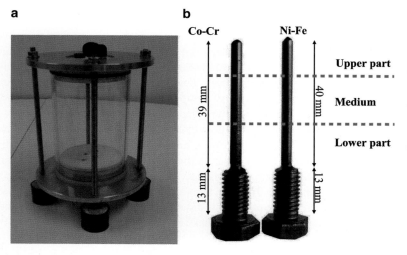

Fig. 8.1 The reaction chamber is a cylinder-shaped element of 100 mm diameter, 150 mm high and 5 mm thick (**a**). The two electrodes presented a height of about 40 mm for the operating part and a diameter of about 3 mm. The threaded portion and the base are 13 and 5 mm long respectively (**b**)

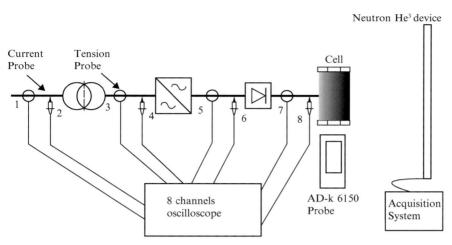

Fig. 8.2 Scheme of the experimental set up adopted and disposition of the measurement equipments employed during the tests

successively filled with the solution. A valve at the top of the cell allows the vapor to escape from the reactor and condense in an external collector. Externally, two circular Inox steel flanges, fastened by means of four threaded ties, hold the Teflon layers. The inferior steel flange of the reactor is connected to four supports isolated from the ground by means of rubber based material. As mentioned before, a direct current passes through the anode and the cathode electrodes, provided by a power circuit connected to the power grid through an electric socket. The components of the circuit are an isolating transformer, an electronic variable transformer (Variac), and a diode bridge linked in series (Fig. 8.2).

8.2.2 Measurement Equipment and Devices

Different physical quantities were measured during the experiments, such as voltage, current, neutron and alpha particle emissions.

Electric current and voltage probes were positioned in different parts of the circuit as shown in Fig. 8.2. The voltage measurements were performed by a differential voltage probe of 100 MHz with a maximum rated voltage of 1400 volts. The current was measured by a Fluke I 310S probe with a maximum rated current of 30 A. Particular attention was paid to the data obtained from the current and voltage probes positioned at the input line powering the reaction chamber (probes 7 and 8 in Fig. 8.2) in order to evaluate the power absorbed by the cell. Current intensity and voltage measurements were also taken by means of a multimeter positioned at the input line. From the turning on to the switching off of the electrolytic cell, current and voltage were found to vary in a range from 3 to 5 A and from 20 to 120 V, respectively. For convenience and clarity, these values are considered as a benchmark to be compared to further measurements which will be reported in future works.

For what concerns the neutron emission measurements, since neutrons are electrically neutral particles, they cannot directly produce ionization in a detector, and therefore cannot be directly detected. This means that neutron detectors must rely upon a conversion process accounting the interaction between an incident neutron and a nucleus, which produces a secondary charged particle. Such charged particle is then detected and the neutron's presence is revealed from it. For an accurate neutron evaluation a He^3 proportional counter was employed. The detector used in the tests is a He^3 type (Xeram, France) with pre-amplification, amplification, and discrimination electronics directly connected to the detector tube. The detector is supplied by a high voltage power (about 1.3 kV) via NIM (Nuclear Instrument Module). The logic output producing the TTL (transistor–transistor logic) pulses is connected to a NIM counter. The logic output of the detector is enabled for analog signals exceeding 300 mV. This discrimination threshold is a consequence of the sensitivity of the He^3 detector to the gamma rays ensuing neutron emission in ordinary nuclear processes. This value has been determined by measuring the analog signal of the detector by means of a Co-60 gamma source. The detector is also calibrated at the factory for the measurement of thermal neutrons; its sensitivity is 65 $cps/n_{thermal}$ (± 10 % declared by the factory), i.e., the flux of thermal neutrons is one thermal neutron/s cm^2, corresponding to a count rate of 65 cps.

For the alpha particle emission, a 6150 AD-k probe with a sealed proportional counter was used, which does not require refilling or flushing from external gas reservoirs. The probe is sensitive to alpha, beta, and gamma radiation. An electronic switch allows for the operating mode "alpha" to detect alpha radiation only, such that in this mode the radiation recognition is very sensitive because the background level is much lower. A removable discriminator plate (stainless steel, 1 mm) distinguishes between beta and gamma radiation detection. An adjustable handle can be locked to the most convenient orientation. During the experiments the 6150 AD-k probe was used in the operating mode alpha to monitor the background level before and after the switching on of the cell.

Finally, before and after the experiments Energy Dispersive X-ray spectroscopy has been performed in order to recognise possible direct evidence of piezonuclear reactions that can take place during the electrolysis. The elemental analyses were performed by a ZEISS Auriga field emission scanning electron microscope (FESEM) equipped with an Oxford INCA energy-dispersive X-ray detector (EDX) with a resolution of 124 eV @ MnKa. The energy used for the analyses was 18 KeV.

8.3   Experimental Results

8.3.1   General Remarks and Preliminary Stage

In Fig. 8.1b, the two electrodes used for the tests are shown. The initial measurement phase implied the use of the Energy Dispersive X-ray spectroscopy (EDX) technique to obtain measurements useful to evaluate the chemical composition of the two electrodes before the experiments. In particular, a series of measures were repeated in three different regions of interest for each electrode in order to obtain a sufficient amount of reliable data. Such regions are the upper, the middle and the lower parts of the single electrode, as reported in Fig. 8.1b.

In Figs. 8.3a and b, the average element concentrations of the electrodes used for the electrolysis are shown. In the initial condition the Ni-Fe electrode (anode) is composed by approximately 44 % in Ni, 30 % in Fe, and 23 % in O. The remaining percentage includes contents of Si, Mn, Ca, Al, K, Na, Mg, Cl, and S, observable only in traces (Fig. 8.3a). On the other hand, the Co-Cr cathode is composed approximately by 44 % in Co, 18 % in Cr, 4 % in Fe, 25 % in O, and traces of

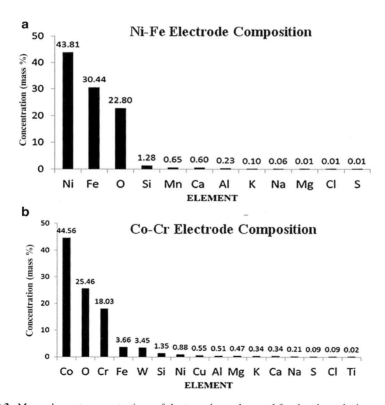

Fig. 8.3 Mean element concentrations of the two electrodes used for the electrolysis

Table 8.1 EDX spectroscopy of the K_2CO_3 salt used for the aqueous solution

Element	Weight%	Atomic%	Compd%	Formula
C	13.02	22.05	47.72	CO_2
K	43.40	22.57	52.28	K_2O
O	43.58	55.38		
Totals	100.00			

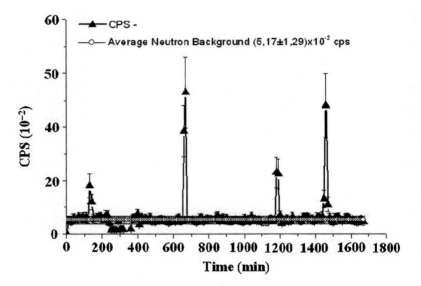

Fig. 8.4 Neutron emission measurements. Emissions up to four and ten times the background level have been observed during the experiments

other elements such as Si, Al, Mg, Na, W, Cu, and S (Fig. 8.3b). Table 8.1 summarizes the results for the compositional analysis conducted on the K_2CO_3 the salt used for the aqueous solution ($K_2CO_3 + H_2O$), where the solute to solvent ratio was approximately 40 g/l.

8.3.2 Neutron and Alpha Particle Detection During the Experiment

Neutron emission measurements performed during the experimental activity are represented in Fig. 8.4. The measurements performed by the He^3 detector were conducted for a total time of about 26 h. The background level was measured for different time spans before and after switching on the reaction chamber. These measurements reported an average neutron background of $(5.17 \pm 1.29 \times 10^{-2})$ cps Furthermore, when the reactor is active, it is possible to observe that after a time span of about 2.5 h (150 min) neutron emissions of about 4 times the background level may be recognized. After 11 h (650 min) from the beginning of the measurements, it is possible to observe a neutron emission level of about one order of magnitude greater than the background level. Similar results were observed after 20 h (1200 min) and 25 h (1500 min) when neutron emissions of about 5 times and 10 times the background were measured, respectively.

In Figs. 8.5a and b, the alpha particle emissions are reported. The data are related to an alpha emission level monitored by means of the 6150 AD-k probe set to the operating mode "alpha". The measurements shown in Fig. 8.5a are referred to the data acquired for a time interval equal to 60 min when the reaction chamber was operating (cell on). The data in Fig 8.5b represent the alpha particle emissions corresponding to the background level and are obtained by measurements acquired, also in this case, for a 60 min (cell off) time interval. From these figures, it can be noticed that the number of counts per second acquired by the probe increased considerably when the electrolytic cell was operating (cell on) (Fig. 8.5a). In addition, the mean values of two alpha emission time series were computed, one when the cell was switched on and the other when the cell was off. The first time series, showed an average alpha emission of about 0.030 c/s (count per second), whereas the second one provided the background emission level in the laboratory with a mean value of about 0.015 c/s. It is evident that the average alpha particle emission during the electrolysis is twice the background level. These results, together with the evidence of neutron emissions reported in Fig. 8.4, are particularly interesting when considering the compositional variation described in the following, and will be useful to corroborate the hypothesis of piezonuclear fission for the chemical elements constituting the electrodes. In Fig. 8.5c, the cumulative curves for the alpha emission counts are reported. It is evident that the total counts value, monitored when the cell was operating (cell on), is approximately twice the value measured for the background level (cell off).

8.3.3 Compositional Analysis of the Electrodes

As reported in the previous section, Energy Dispersive X-Ray Spectroscopy was performed in order to recognise possible direct evidence of piezonuclear reactions taking place during the electrolysis. The measurement precision is in the order of magnitude of 0.1 %. In Fig. 8.6a and b two images of the Co-Cr electrode surface respectively before the experiment and after 32 h from the beginning of it are reported. It is shown that the electrode after several operating hours presented micro-cracks and cracks visible on its external surface (see Fig. 8.6b).

The experimental activity was developed in three different phases in order to investigate possible compositional variations on the electrode surface. A first analysis was carried out to evaluate the composition of the electrodes before they underwent the electrolysis experiment (0 h), (see Table 8.2). The second analysis was conducted after an initial operating time of the electrolytic cell of about 4 h (see

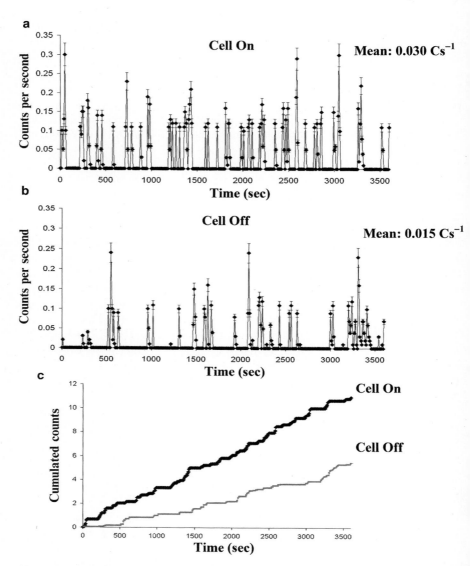

Fig. 8.5 Data acquired for a time interval equal to 60 min when the reaction chamber was operating (**a**). The alpha particle emissions corresponding to the background level (**b**). Cumulative curves for the alpha emissions (**c**)

Table 8.2). After this, a third and a fourth step analyses were performed. For these two steps, the cell operated for 28 and 6 h respectively, corresponding to a cumulative working time of 32 (4 h + 28 h) and 38 h (4 h + 28 h + 6 h) (see Table 8.2). In the case of the Ni-Fe electrode, the resulting mean concentrations

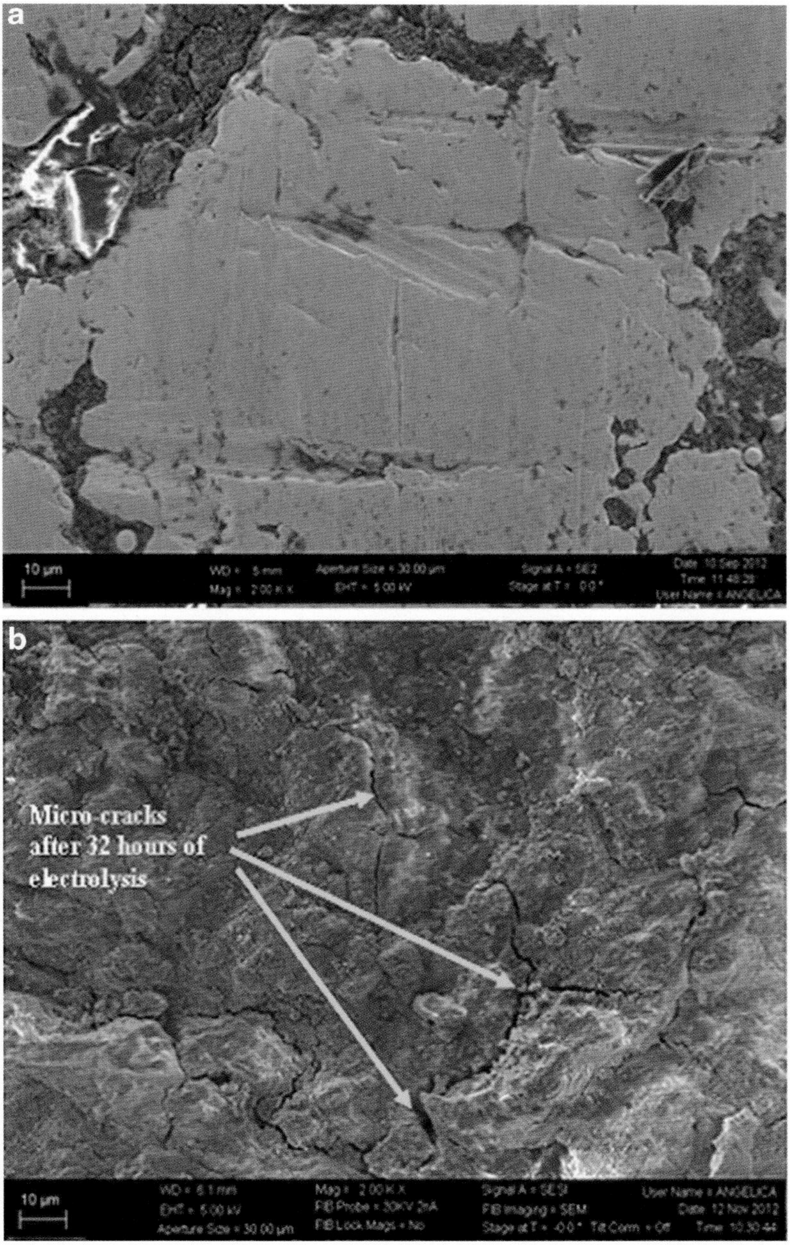

Fig. 8.6 Image of the Co-Cr electrode surface before the experiment (**a**). The electrode after 32 operating hours presented cracks and micro-cracks visible on the external surfaces (**b**)

Table 8.2 Ni-Fe electrode, element concentration before the experiment, after 4, 32 and 38 h of the test

	Ni (%)	Si (%)	Mg (%)	Fe (%)	Cr (%)
Before the experiment	43.9	1.1	0.1	30.5	–
After 4 h	43.6	1.1	0.4	30.7	–
After 32 h	35.2	5.0	0.2	27.9	–
After 38 h	35.3	1.5	4.8	27.3	3.0

*The values reported for the mass % of each element are referred to the mean value of all the effectuated measurements

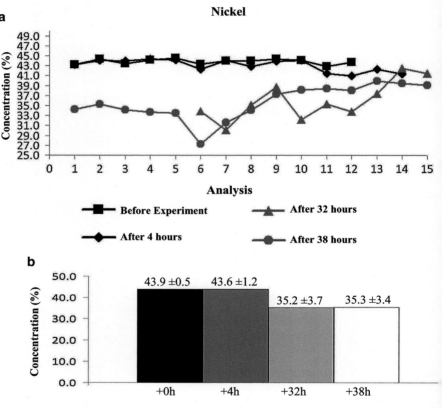

Fig. 8.7 Ni concentration before the experiment, after 4, 32 and 38 h (**a**). The average values of Ni concentration change from a mass percentage of 43.9 % at the beginning of the experiment to 35.2 % and 35.3 % after 32 and 38 h respectively (**b**)

of Ni, Si, Mg, Fe, and Cr are reported for each step, see Table 8.2. In Figs. 8.7, 8.8, 8.9, 8.10, and 8.11, the EDX measurements for each element are reported considering the four steps previously mentioned (Figs. 8.7a, 8.8a, 8.9a, 8.10a and 8.11a). At the same time, the evolution of the mean values of each time series along with

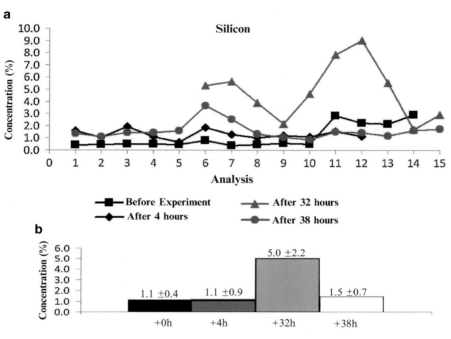

Fig. 8.8 Si concentration before the experiment, after 4, 32 and 38 h (**a**). The mean values of Si concentration change from a mass percentage of 1.1 % at the beginning of the experiment to 5.0 % and 1.5 % after 32 and 38 h respectively (**b**)

their respective standard deviations, corresponding to 0, 4, 32, and 38 h, are reported by histograms (Figs. 8.7b, 8.8b, 8.9b, 8.10b and 8.11b). After 38 h, the appearance of Cr, before absent, was detected as reported in Table 8.2 and in Fig. 8.11a. In particular, the Ni concentration showed a total average decrease of 8.6 % from 43.9 % to 35.3 % after 38 h (see Table 8.2 and Figs. 8.7a and b). This Ni depletion is one fifth of the initial Ni concentration. A mean increment in Si concentration after 32 h of 3.9 % and an average increment in Mg concentration after 38 h, starting from 0.1 % up to 4.8 %, can be observed from the data reported in Table 8.2, Figs. 8.8 and 8.9. Similar considerations may be done also for Fe and Cr concentrations. The average Fe content decreased of 3.2 %, changing from 30.5 % to 27.3 % at the end of the experiment (see Table 8.2 and Fig. 8.10). On the other hand, the Cr concentration appeared only in the last phase with an appreciable increase of 3.0 % (see Table 8.2 and Fig. 8.11). Such decrements in Ni and Fe seem to be almost perfectly counterbalanced by the increments in other elements: Si, Mg, and Cr. In particular, since the analysis excluded Ni content variations on the other electrode, the balance: Ni $(-8.6\ \%) = $ Si $(+3.9\ \%) + $ Mg $(+4.7\ \%)$ can be reasonably explained only by the following symmetrical piezonuclear fissions:

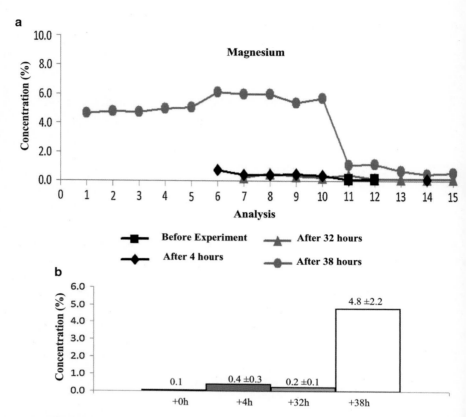

Fig. 8.9 Mg concentration before the experiment, after 4, 32 and 38 h (**a**). The mean values of Mg concentration change from a mass percentage of 0.1 % and 0.4 % at the beginning of the experiment to 4.8 % after 38 h (**b**)

$$Ni_{28}^{58} \rightarrow 2Si_{14}^{28} + 2 \text{ neutrons} \tag{8.1}$$

$$Ni_{28}^{58} \rightarrow 2Mg_{12}^{24} + 2He_{2}^{4} + 2 \text{ neutrons} \tag{8.2}$$

At the same time, the balance Fe (-3.2 %) $\cong$ Cr ($+3.0$ %) may be explained by the reaction:

$$Fe_{26}^{56} \rightarrow Cr_{24}^{52} + He_{2}^{4} \tag{8.3}$$

It is very interesting to notice that reactions (8.1) and (8.2) imply neutron emissions, as well as reactions (8.2) and (8.3) imply the emissions of alpha particles.

As far as the Co-Cr electrode is concerned, it is possible to observe variations even more evident in the concentrations of the most abundant constituting elements. In particular, the average Co concentration decreased by 23.5 %, from an initial percentage of 44.1 % to a concentration of 20.6 % after 32 h (see Table 8.3

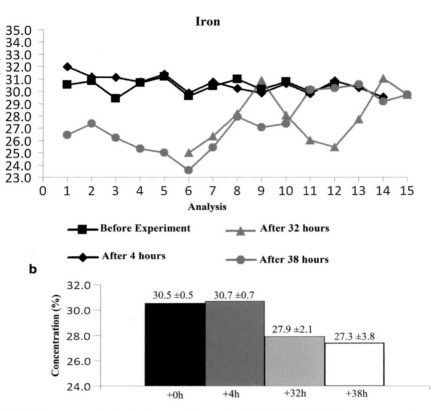

Fig. 8.10 Fe concentration before the experiment, after 4, 32 and 38 h (**a**). The mean values of Fe concentration change from a mass percentage of 30.5 % at the beginning of the experiment to 27.9 % and 27.3 % after 32 and 38 h respectively (**b**)

and Figs. 8.12a and b). At the same time an increment of 23.2 % can be observed in the Fe content after 32 h, changing from 3.1 % before the experiment to 26.3 % at the end of the second phase (see Table 8.3 and Figs. 8.13a and b). It is rather impressive that the decrease in Co and the increase in Fe are almost the same: Co (-23.5 %) $\cong$ Fe ($+23.2$ %). According to the compositional analysis on the Ni-Fe electrode no Co concentration was found that could lead us to consider the chemical migration to explain its depletion in the Co-Cr electrode, thus the following reaction seems the most reasonable possibility:

$$Co_{27}^{59} \rightarrow Fe_{26}^{56} + H_1^1 + 2 \text{ neutrons} \qquad (8.4)$$

In Table 8.3 and in Figs. 8.14 and 8.15, Cr and K concentrations are reported for the different phases of the experiment. The K increase by about 12.4 % after 32 h may be only partially counterbalanced by the decrease in Cr (8.1 %) according to

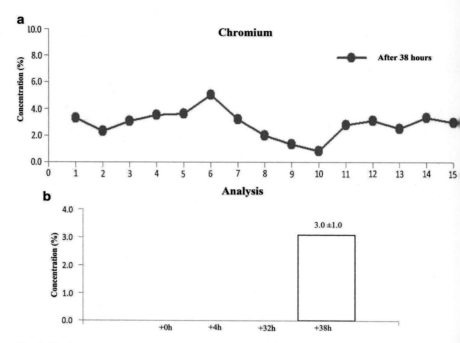

Fig. 8.11 Cr concentration after 38 h (**a**). The mean value of Cr concentration changes from 0 % (absence of this element) to a mass percentage of 3.0 % after 38 h (**b**)

Table 8.3 Co-Cr electrode, element concentration before the experiment, after 4, 32 and 38 h of the test

	Co (%)	Fe (%)	Cr (%)	K (%)
Before the experiment:	44.1	3.1	17.8	0.5
After 4 h	43.7	1.6	17.8	2.2
After 32 h	20.6	26.3	9.7	12.9
After 38 h	34.4	6.6	5.1	4.4

*The values reported for the mass % of each element are referred as the mean value of all the effectuated measurements

reaction (8.5). The remaining increment of K (4.3 %) could be considered as an effect of the K_2CO_3 aqueous solution deposition at the end of the third step.

Considering these variations also the following piezonuclear reaction, involving Cr as the starting element and K as the resultant could be considered:

$$Cr_{24}^{52} \rightarrow K_{19}^{39} + H_1^1 + 2He_2^4 + 4 \text{ neutrons} \qquad (8.5)$$

Also in this case, it is remarkable that both reactions (8.4) and (8.5) imply neutron emissions, whereas reaction (8.5) implies also the emission of alpha particles.

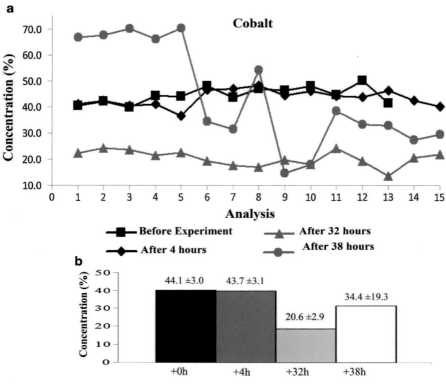

Fig. 8.12 Co concentration before the experiment, after 4, 32 and 38 h (**a**). The mean values of Co concentration change from a mass percentage of 44.1 % at the beginning of the experiment to 20.6 % and 34.4 % after 32 and 38 h respectively (**b**)

It is important to consider that the balances reported, for the Ni-Fe electrode (reactions 8.1, 8.2 and 8.3) and for the Co-Cr electrode (reactions 8.4 and 8.5), were obtained considering the values of the second or third step corresponding to the largest variation for each element concentration (see Tables. 8.2 and 8.3). Additional variations observed for some of these elements, such as Si, Co and Fe, between the second and the third step, may be explained considering other possible secondary piezonuclear reactions occurring on the electrode surface. Further efforts should be devoted to evaluate the evidence of these secondary fissions. All these results suggest that during the gas loading, performed with hydrogen or deuterium, the host lattice is subjected to mechanical damaging and fracturing due to atoms absorption and penetration. Evidences of a diffused cracking were identified also on the electrode surface after the experiments (see Fig. 8.6b). According to this, we argue that the hydrogen, favoring the crack formation and propagation in the metal, comes from the electrolysis of water. In fact, being the electrodes immersed in a liquid solution, their surface is exposed to the formation of gaseous hydrogen due to the decomposition of water caused by the current passage.

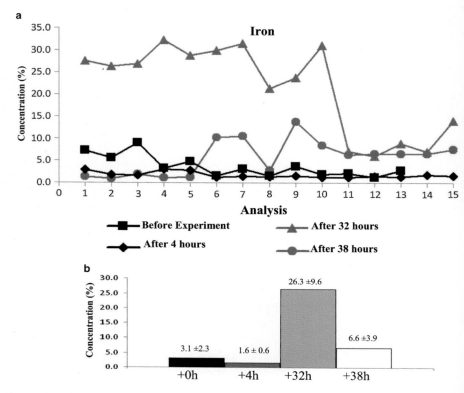

Fig. 8.13 Fe concentration before the experiment after 4, 32 and 38 h (**a**). The mean values of Fe concentration change from a mass percentage of 3.1 % at the beginning of the experiment to 26.3 % and 6.6 % after 32 and 38 h respectively (**b**)

8.4 Conclusions

Neutron emissions up to one order of magnitude higher than the background level were observed during the operating time of an electrolytic cell. In particular, after a time span of about 3 h, neutron emissions of about four times the background level were measured. After 11 h, it was possible to observe neutron emissions of about one order of magnitude greater than the background level. Similar results were observed after 20 and 25 h.

When the cell was switched on, the average alpha particle emission was about 0.030 c/s for 1 h of measurement; this value corresponds to an alpha emission level of twice the background measured in the laboratory before and after the experiment (0.015 c/s).

By the EDX analysis performed on the two electrodes in three successive steps, significant compositional variations could be recorded. In general, the decrements in Ni and Fe in the Ni-Fe electrode seem to be almost perfectly counterbalanced by

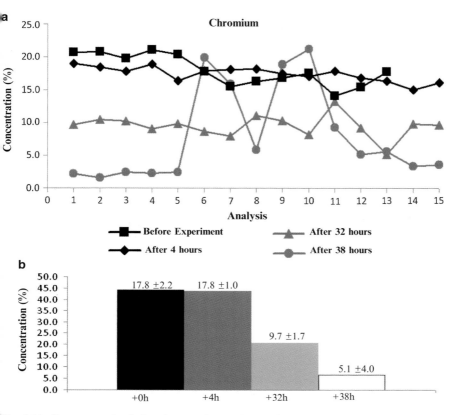

Fig. 8.14 Cr concentration before the experiment after 4, 32 and 38 h (**a**). The Cr concentration changes from a mass percentage of 17.8–9.7 % and 5.1 % after 32 and 38 h (**b**)

the increments in lighter elements: Si, Mg, and Cr. In fact, the balance Ni $(-8.6 \%) \cong Si (+3.9 \%) + Mg (+4.7 \%)$ is satisfied by reactions (8.1) and (8.2). Specifically, Si could have also undergone further reactions, which would explain its drop in concentration observed at the third step. At the same time, the balance Fe $(-3.2 \%) \cong Cr (+3.0 \%)$ may be explained considering reaction (8.3).

As far as the Co-Cr electrode is concerned, the Co decrease is almost perfectly counterbalanced by the Fe increase: Co $(-23.5 \%) \cong Fe (+23.2 \%)$. This last evidence, which is really impressive considering the mass percentage involved, seems to be explainable only considering reaction (8.4). Finally, the Cr decrease and the K increase may be explained taking into account reaction (8.5) and the solution deposition. In particular, the K increase by about 12.4 % may be only partially counterbalanced by the decrease in Cr (8.1 %) according to reaction (8.5). The remaining increment in K (4.3 %) could be considered as an effect of the K_2CO_3 aqueous solution deposition at the end of the third step.

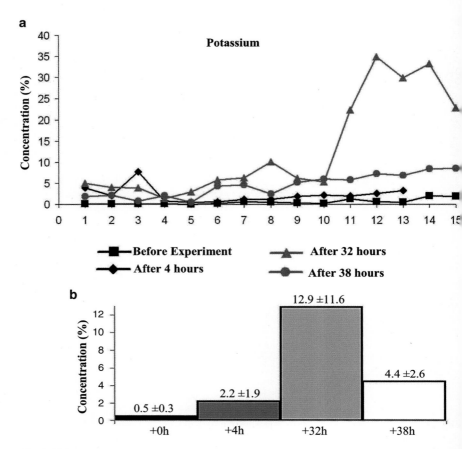

Fig. 8.15 K concentration before the experiment after 4, 32 and 38 h (**a**). The K concentration changes from 0.5 % and 2.2 % after 4 h to 12.9 % and 4.4 % after 32 and 38 h (**b**)

Electrochemical reasons cannot explain the variations observed. Elements such Ni and Co have shown no traces on the surface of the other electrode and could no be found dissolved in the solution. Also the so-called Elettromigration, having a reduced effect due to the size of the electrode, cannot be considered for the explanation of compositional variations of the order of tens of percentage points.

The Chemical variations and the energy emissions may be accounted for direct and indirect evidence of mechano-nuclear fission reactions correlated to microcrack formation and propagation due to hydrogen embrittlement. According to this interpretation of the so-called Cold Nuclear Fusion, hydrogen, which is imputed to favor the crack formation and propagation in the electrodes, comes from the electrolysis of water. Being the electrodes immersed in a liquid solution, the metal surface is exposed to the formation of gaseous hydrogen due to the decomposition of water molecule caused by the current passage. In addition, the high current density contributes to the formation and penetration of hydrogen into the metal. The so-called Cold Nuclear Fusion, interpreted under the light of hydrogen embrittlement, may be explained by piezonuclear fission reactions occurring in the host metal, rather than by nuclear fusion of hydrogen isotopes forced into the metal lattice.

Aknowledgements Dr. A. Sardi and Dr. F. Durbiano are gratefully acknowledged for their help during the experimental set-up definition and for the assistance given during the measurements. Also Dr. A. Chiodoni is warmly acknowledged for the EDX spectroscopy analyses.

References

1. Borghi DC, Giori DC, Dall'Olio A (1992) Experimental evidence on the emission of neutrons from cold hydrogen plasma. In: Proceedings of the international workshop on few-body problems in low-energy physics, Alma-Ata, Kazakhstan, Unpublished Communication (1957). Comunicacao n. 25 do CENUFPE, Recife Brazil (1971), pp 147–154
2. Diebner K (1962) Fusionsprozesse mit Hilfe konvergenter Stosswellen – einige aeltere und neuere Versuche und Ueberlegungen. Kerntechnik 3:89–93
3. Kaliski S (1978) Bi-conical system of concentric explosive compression of D-T. J Tech Phys 19:283–289
4. Winterberg F (1984) Autocatalytic fusion – fission implosions. Atomenergie-Kerntechnik 44:146
5. Derjaguin BV et al (1989) Titanium fracture yields neutrons? Nature 34:492
6. Fleischmann M, Pons S, Hawkins M (1989) Electrochemically induced nuclear fusion of deuterium. J Electroanal Chem 261:301
7. Bockris JM, Lin GH, Kainthla RC, Packham NJC, Velev O (1990) Does tritium form at electrodes by nuclear reactions? The first annual conference on cold fusion. University of Utah Research Park/National Cold Fusion Institute, Salt Lake City
8. Preparata G (1991) Some theories of cold fusion: a review. Fusion Techol 20:82
9. Preparata G (1991) A new look at solid-state fractures, particle emissions and "cold" nuclear fusion. Il Nuovo Cimento 104A:1259–1263
10. Mills RL, Kneizys P (1991) Excess heat production by the electrolysis of an aqueous potassium carbonate electrolyte and the implications for cold fusion. Fusion Technol 20:65
11. Notoya R, Enyo M (1992) Excess heat production during electrolysis of H_2O on Ni, Au, Ag and Sn electrodes in alkaline media. In: Proceedings of the third international conference on cold fusion, Nagoya Japan, Universal Academy Press, Tokyo
12. Miles MH, Hollins RA, Bush BF, Lagowski JJ, Miles RE (1993) Correlation of excess power and Helium production during D_2O and H_2O electrolysis using palladium cathodes. J. Electroanal Chem 346:99–117
13. Bush RT, Eagleton RD (1993) Calorimetric studies for several light water electrolytic cells with nickel fibrex cathodes and electrolytes with alkali salts of potassium, rubidium, and cesium. Fourth international conference on cold fusion. Lahaina, Maui. Electric Power Research Institute 3412 Hillview Ave., Palo Alto 94304. 13
14. Fleischmann M, Pons S, Preparata G (1994) Possible theories of cold fusion. Nuovo Cimento Soc Ital Fis A 107:143
15. Szpak S, Mosier-Boss PA, Smith JJ (1994) Deuterium uptake during Pd-D codeposition. J Electroanal Chem 379:121
16. Sundaresan R, Bockris JOM (1994) Anomalous reactions during arcing between carbon rods in water. Fusion Technol 26:261

17. Arata Y, Zhang Y (1995) Achievement of solid-state plasma fusion ("cold-fusion"). Proc Jpn Acad 71(B):304–309
18. Ohmori T, Mizuno T, Enyo M (1996) Isotopic distributions of heavy metal elements produced during the light water electrolysis on gold electrodes. J New Energy 1(3):90–99
19. Monti RA (1996) Low energy nuclear reactions: experimental evidence for the alpha extended model of the atom. J New Energy 1(3):131
20. Monti RA (1998) Nuclear transmutation processes of lead, silver, thorium uranium. The seventh international conference on gold fusion. ENECO, Vancouver/Salt Lake City
21. Ohmori T, Mizuno T (1998) Strong excess energy evolution, new element production, and electromagnetic wave and/or neutron emission in light water electrolysis with a tungsten cathode. Infinite Energy 20:14–17
22. Mizuno T (1998) Nuclear transmutation: the reality of cold fusion. Infinite Energy Press, New Hampshire
23. Little SR, Puthoff HE, Little ME (1998) Search for excess heat from a Pt electrode discharge in K_2CO_3-H_2O and K_2CO_3-D_2O electrolytes. Infinite Energy 5:34
24. Ohmori T, Mizuno T (2000) Nuclear transmutation reaction caused by light water electrolysis on tungsten cathode under incandescent conditions. J New Energy 4(4):66–78
25. Ransford HE (1999) Non-Stellar nucleosynthesis: transition metal production by DC plasma-discharge electrolysis using carbon electrodes in a non-metallic cell. Infinite Energy 4 (23):16–22
26. Storms E (2000) Excess power production from platinum cathodes using the Pons-Fleischmann effect. Eigth international conference on cold fusion. Lerici (La Spezia). Italian Physical Society, Bologna, pp 55–61
27. Storms E (2007) Science of low energy nuclear reaction: a comprehensive compilation of evidence and explanations about cold fusion. World Scientific Publishing, Singapore
28. Mizuno T et al (2000) Production of heat during plasma electrolysis. Jpn J Appl Phys A 39:6055
29. Warner J, Dash J and Frantz S (2002) Electrolysis of D2O with titanium cathodes: enhancement of excess heat and further evidence of possible transmutation. The 9th international conference on cold fusion, Tsinghua University, Beijing, p 404
30. Fujii MF et al (2002) Neutron emission from fracture of piezoelectric materials in deuterium atmosphere. Jpn J Appl Phys 41:2115–2119
31. Mosier-Boss PA et al (2007) Use of CR-39 in Pd/D co-deposition experiments. Eur Phys J Appl Phys 40:293–303
32. Swartz M (2008) Three physical regions of anomalous activity in deuterated palladium. Infinite Energy 14:19–31
33. Mosier-Boss PA et al (2010) Comparison of Pd/D co-deposition and DT neutron generated triple tracks observed in CR-39 detectors. Eur Phys J Appl Phys 51(2):20901–20911
34. Kanarev M, Mizuno T (2002) Cold fusion by plasma electrolysis of water. J Theoretics 5:1–7
35. Cardone F, Mignani R (2003) Possible observation of transformation of chemical elements in cavitated water. Int J Mod Phys B 17:307–317
36. Cardone F, Cherubini G, Petrucci A (2009) Piezonuclear neutrons. Phys Lett A 373:862–866
37. Carpinteri A, Cardone F, Lacidogna G (2009) Piezonuclear neutrons from brittle fracture: early results of mechanical compression tests. Strain 45:332–339. Atti dell'Accademia delle Scienze di Torino 33:27–42
38. Cardone F, Carpinteri A, Lacidogna G (2009) Piezonuclear neutrons from fracturing of inert solids. Phys Lett A 373:4158–4163
39. Carpinteri A, Cardone F, Lacidogna G (2010) Energy emissions from failure phenomena: mechanical, electromagnetic, nuclear. Exp Mech 50:1235–1243
40. Carpinteri A, Lacidogna G, Manuello A, Borla O (2012) Piezonuclear fission reactions: evidences from microchemical analysis, neutron emission, and geological transformation. Rock Mech Rock Eng 45:445–459

41. Cardone F, Mignani R, Petrucci A (2012) Piezonuclear reactions. J Adv Phys 1:3
42. Albertini G, Calbucci V, Cardone F, Petrucci A, Ridolfi F (2014) Chemical changes induced by ultrasounds in iron. Appl. Physics A 114(4):1233–1246. doi:10.1007/s00339-013-7876-z
43. Petrucci A, Mignani R, Cardone F (2011) Comparison between piezonuclear reactions and CMNS phenomenology. In: Violante V, Sarto F (eds) Proceeding of the 15th international conference on condensed matter nuclear science (Part 1), Rome, 5–9 Oct 2009, p 246
44. Carpinteri A, Lacidogna G, Manuello A, Borla O (2013) Piezonuclear fission reactions from earthquakes and brittle rocks failure: evidence of neutron emission and nonradioactive product elements. Exp Mech 53:345–365
45. Milne I, Ritchie RO, Karihaloo B (2003) Comprehensive structural integrity: fracture of materials from nano to macro. Elsevier, Amsterdam, pp 31–33, Volume 6, Chapter 6.02
46. Liebowitz H (1971) Fracture an advanced treatise. Academic Press, New York/San Francisco/London

Chapter 9
Cold Nuclear Fusion Explained by Hydrogen Embrittlement and Piezonuclear Fissions in Metallic Electrodes: Part II: Pd and Ni Electrodes

Alberto Carpinteri, Oscar Borla, Alessandro Goi, Salvatore Guastella, Amedeo Manuello, and Diego Veneziano

Abstract Recent experiments provided evidence of piezonuclear reactions occurring in condensed matter during electrolysis. These experiments were characterized by significant neutron and alpha particle emissions, together with appreciable variations in the chemical composition at the electrode surfaces. A mechanical reason for the so-called Cold Nuclear Fusion was recently proposed by the authors. The hydrogen embrittlement due to H atoms produced by the electrolysis plays an essential role for the observed micro-cracking in the electrode host metals (Pd and Ni). Consequently, our hypothesis is that piezonuclear fission reactions may occur in correspondence to the micro-crack formation or propagation. In order to confirm the early results obtained by the Ni-Fe and Co-Cr electrodes and presented in the companion paper (Part I), electrolytic tests have been conducted using 100 % Pd at the cathode and 90 % Ni at the anode. As a result, relevant compositional changes and the appearance of elements previously absent have been observed on the Pd and Ni electrodes after the experiments, as well as significant neutron emissions. The most relevant process emerging from the experiment is the primary fission of palladium into iron and calcium. Then, secondary fissions of both the products as well as of nickel in the other electrode appear in turn producing oxygen atoms, alpha particles, and neutrons.

Keywords Hydrogen embrittlement • Cold nuclear fusion • Electrolysis • Piezonuclear reactions • Neutron emission • Palladium • Nickel

A. Carpinteri (✉) • O. Borla • A. Manuello • D. Veneziano
Department of Structural, Geotechnical and Building Engineering,
Politecnico di Torino, Corso Duca degli Abruzzi 24, 10129 Torino, Italy
e-mail: alberto.carpinteri@polito.it

A. Goi
Private experimentalist Varese, Italy

S. Guastella
Department of Applied Science and Technology,
Politecnico di Torino, Corso Duca degli Abruzzi 24, 10129 Torino, Italy

© Springer International Publishing Switzerland 2015
A. Carpinteri et al. (eds.), *Acoustic, Electromagnetic, Neutron Emissions from Fracture and Earthquakes*, DOI 10.1007/978-3-319-16955-2_9

9.1 Introduction

Several evidences of anomalous nuclear reactions occurring in condensed matter were observed by different authors [1–34]. These experiments are characterized by extra-heat generation, neutron emission, and alpha particle detection. Some of these studies, using electrolytic devices, reported also significant evidences of compositional variation after micro-cracking of the electrodes [35–40].

As reported in the companion paper (Part I), in 1998, Mizuno presented the results of the measurements conducted by means of neutron detectors and compositional analysis techniques related to different electrolytic experiments. On the other hand, as shown by most of the articles devoted to Cold Nuclear Fusion, one of the principal features is the appearance of micro-cracks on electrode surfaces after the tests [26, 27]. Such evidence might be directly correlated to hydrogen embrittlement of the material composing the metal electrodes (Pd, Ni, Fe, Ti, etc.). This phenomenon, well-known in Metallurgy and Fracture Mechanics, characterizes metals during forming or finishing operations [41, 42]. In order to confirm the early results obtained by Ni-Fe and Co-Cr electrodes [43–45], reported also in Part I [45], electrolytic tests have been conducted using 100 % Pd at the cathode and a Ni-Fe alloy (91 % of Ni) at the anode. In the present study, the host metal matrix is subjected to mechanical damaging and fracturing due to hydrogen atoms penetrating into the lattice structure and forcing it, during the gas loading. Hydrogen effects are largely studied, especially in metal alloys, where the hydrogen absorption is particularly high. The hydrogen atoms generate an internal stress that lowers the fracture resistance of the metal, so that brittle crack formation or propagation can occur with a hydrogen partial pressure below 1 atm [41, 42]. These results give an important confirmation about the hypothesis proposed by the authors and reported in previous papers on electrolysis with Ni-Fe and Co-Cr electrodes [43, 44].

9.2 Experimental Set-Up

The experimental set-up adopted during the tests is the same as described in the companion paper (Part I) [45]. Similarly to the previous experiments, the two metallic electrodes were connected to a source of direct current: a Ni-Fe electrode as the positive pole (anode), and a Pd electrode as the negative pole (cathode). The reaction chamber is the same employed by the authors for the previous test, the current intensity and the neutron emissions were analysed according to the same procedure adopted in the case of the Ni-Fe and Co-Cr electrodes.

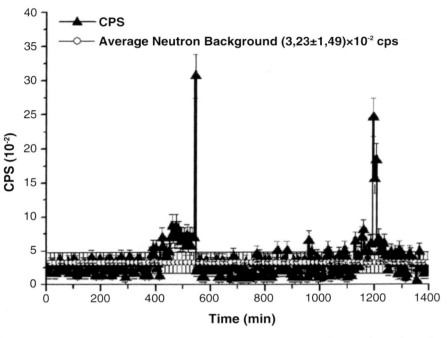

Fig. 9.1 Neutron emission measurements. Emissions between three times and ten times the background level have been observed during the experiment

9.3   Neutron Emission Measurements

Neutron emission measurements performed during the experimental activity are represented in Fig. 9.1. The measurements performed by the He^3 detector were conducted for a total time of 24 h. The background level was measured for different time intervals before and after switching on the reaction chamber. These measurements provided an average neutron background of $(3.23 \pm 1.49) \times 10^{-2}$ cps. Furthermore, when the reaction chamber was active, it was possible to observe that, after a time interval of about 7.5 h (460 min), neutron emissions of about 3 times the background level were detected. After 9 h (545 min) from the beginning of the measurements, it was possible to observe a neutron emission level of about one order of magnitude greater than the background level. Similar results were observed after 20 h (1200 min), when neutron emissions up to seven times the background were measured.

9.4   Compositional Analysis of the PD Electrode

In the present section, the chemical compositions before and after the experiment will be taken into account, as well as the concentrations measured for each element identified on the surfaces of the two electrodes (see Tables 9.1 and 9.2). The measurement precision is in the order of magnitude of 0.1 %.

Table 9.1 Element concentrations before and after the electrolysis (Pd electrode)

	Mean values						
	Pd	Fe	Ca	O	Mg	K	Si
After 0 hours (%)	100.0	0.0	0.0	0.0	0.0	0.0	0.0
After 20 hours (%)	71.3	2.0	0.2	18.5	1.0	1.5	1.1

Table 9.2 Element concentrations before and after the electrolysis (Ni electrode)

	Mean values				
	Ni	O	Si	Fe	Al
After 0 hours (%)	91.6	2.0	0.3	2.4	0.0
After 20 hours (%)	68.5	21.5	1.1	0.4	1.8

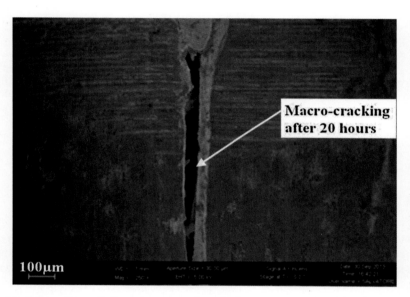

Macro-cracking after 20 hours

100μm

Fig. 9.2 Image of the palladium electrode surface: the fracture presented a width of about 40 μm

In particular, considering the neutron emission measurements and according to the hydrogen embrittlement hypothesis suggested by Carpinteri et al. [43, 44], the presence of micro-cracks and macro-cracks on the electrode surface (Fig. 9.2) is accounted in the mechanical interpretation of the phenomena. These evidences are particularly strong in the case of the Pd electrode, where a macroscopic fracture took place during the test. The fracture presented a width of about 40 μm observable also at naked eyes (see Fig. 9.2).

Considering the average decrement in Pd (−28.6 %), reported in Table 9.1, a first-generation fission can be assumed:

$$Pd_{46}^{106} \rightarrow Ca_{20}^{40} + Fe_{26}^{56} + 10 \text{ neutrons} \tag{9.1}$$

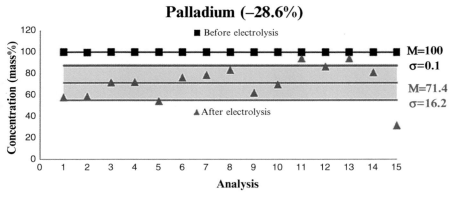

Fig. 9.3 Pd concentrations measured on 15 different spots of the electrode surface, before (*black*) and after (*red*) the electrolysis; the average concentration M and the corresponding stand. dev. σ are reported

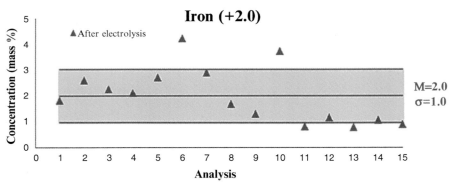

Fig. 9.4 Fe concentrations measured on the electrode surface after the electrolysis; iron has not been detected before the experiment

According to reaction (9.1), the Pd decrement is counterbalanced by the Ca and Fe increments in the following proportions: 10.8 % and 15.1 %, respectively. These variations are accompanied by a neutron emission corresponding to the remaining 2.7 % of the mass concentration (see also Figs. 9.3, 9.4, and 9.5).

The whole iron increment, according to reaction (9.1), could be entirely taken as the starting element for the production of other elements. Hence, a second hypothesis can be considered involving Fe as starting element and O as the product, together with alpha particle and neutron emissions (see Fig. 9.6):

$$Fe_{26}^{56} \rightarrow 3O_8^{16} + He_2^4 + 4 \ \text{neutrons} \tag{9.2}$$

According to reaction (9.2), the iron decrement produces 12.9 % of oxygen with alpha particle (He) and neutron emissions. The total measured increment in oxygen

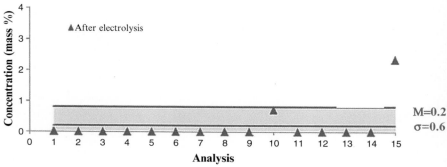

Fig. 9.5 Traces of Ca concentrations have been detected after the electrolysis

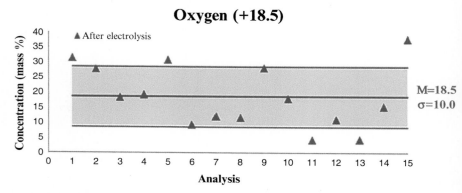

Fig. 9.6 The presence of the oxygen is remarkable after the experiments

after the experiment is equal to 18.5 %, (see Table 9.1). This quantity seems to be only partially explained by reaction (9.2). The remaining 5.6 % of O concentration could be explained considering other reactions involving Ca, produced in reaction (9.1), as the starting element:

$$Ca_{20}^{40} \rightarrow 2O_8^{16} + 2He_2^4 \tag{9.3}$$

$$Ca_{20}^{40} \rightarrow O_8^{16} + Mg_{12}^{24} \tag{9.4}$$

From reaction (9.3) we can consider a decrement in Ca equal to 5.9 %. This decrement gives an increment of 4.7 % in O and 0.6 in He. On the other hand, from reaction (9.4), we obtain a further decrease in Ca concentration of 1.6 %, as well as the formation of 1.0 % of Mg and 0.6 % of O (Fig. 9.7).

The calculated O increase of 18.2 % is very close to the experimental value of 18.5 %. At the same time, the Mg increment observed after the experiment can be explained by reaction (9.4), see also Table 9.1. According to reactions (9.3) and

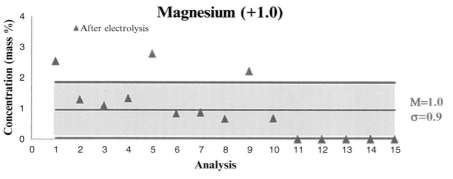

Fig. 9.7 The magnesium presence is evident after the experiment, while there was no trace of it before

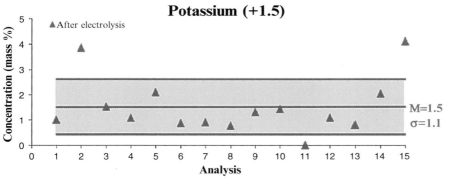

Fig. 9.8 Potassium evidences have been detected only after the experiment

(9.4), the following balances may be considered: Ca $(-5.9\ \%) = $ O $(+4.7\ \%) +$ He $(+1.2\ \%)$, and Ca $(-1.6\ \%) = $ O $(+0.6\ \%) +$ Mg $(+1.0\ \%)$. Taking into account the same reactions, and considering the Ca increment coming from reaction (9.1) $(10.8\ \%)$, a concentration of 3.3 % remains to be matched. To this aim, it is possible to take into account additional reactions involving Ca as starting element and Si, K, and C as the products (Figs. 9.8 and 9.9):

$$Ca_{20}^{40} \rightarrow K_{19}^{39} + H_1^1 \tag{9.5}$$

$$Ca_{20}^{40} \rightarrow C_6^{12} + Si_{14}^{28} \tag{9.6}$$

From these reactions, the following balances may be considered: Ca $(-1.5\ \%) = $ K $(+1.5\ \%)$, and Ca $(-1.6\ \%) = $ C $(+0.5\ \%) +$ Si $(+1.1\ \%)$. Considering the experimental residual 0.2 % of Ca, and the previously calculated residual 3.3 % of Ca, the previous two reactions provide a complete matching.

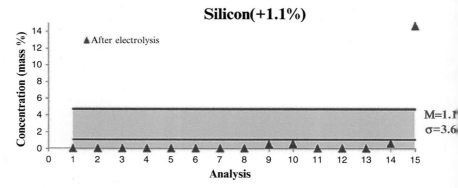

Fig. 9.9 EDS measures show evident traces of Silicon only after electrolysis

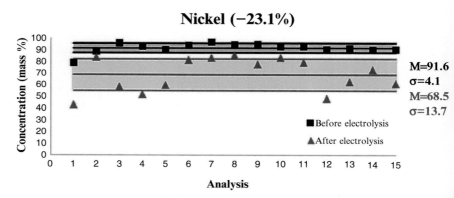

Fig. 9.10 Nickel electrode: the series of the Ni concentrations measured before and after 20 h of electrolysis are showed. An average variation between the two series of about 23.1 % may be considered

9.5 Compositional Analysis of the Ni Electrode

Let us consider the nickel electrode. Table 9.2 summarizes the concentration variations after the electrolysis. Nickel diminishes by 23.1 %, whereas the most relevant positive variation is that of oxygen (+19.5 %). It is worth observing that the average concentration decrement in Fe (−2.0 %) is comparable to the average increment in Al (see Table 9.2). The Figs. 9.10 and 9.11 show the set of concentrations measured before and after the experiment for Ni and O.

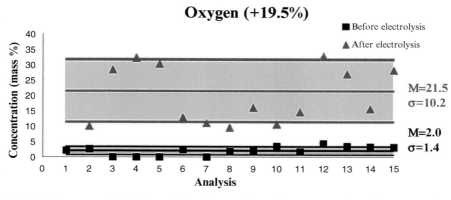

Fig. 9.11 O concentration before and after the experiment. The average values of O concentration change from a mass percentage of 2.0 % at the beginning of the experiment to 21.5 %

On the basis of the piezonuclear reaction conjecture, we could assume the oxygen average variation as a nuclear effect caused by the following reaction:

$$\text{Ni}_{28}^{59} \rightarrow 3\text{O}_8^{16} + 2\text{He}_2^4 + 3 \text{ neutrons} \tag{9.7}$$

A second hypothesis could be considered for the Al average variation, which is consistent with the following reaction:

$$\text{Fe}_{26}^{56} \rightarrow 2\text{Al}_{13}^{27} + 2 \text{ neutrons} \tag{9.8}$$

A third hypothesis could be made considering the silicon average variation:

$$\text{Ni}_{28}^{59} \rightarrow 2\text{Si}_{14}^{28} + 3 \text{ neutrons} \tag{9.9}$$

Reactions (9.7), (9.8), and (9.9) imply emissions of neutrons, which provide a great support to the hypotheses based on piezonuclear reactions. The hypothesis is that an average decrement of 22.1 % in Nickel underwent a reaction producing at least 18.0 % of oxygen together with alpha and neutron emissions. Secondly, another average decrease in Ni of 1 % could have gone into silicon (+0.9 %) and more neutrons. Thirdly, an average depletion of 2 % in Fe would have produced about 1.9 % of Al accompanied by neutron emissions (+0.1 %). The calculated O increment of 18.0 % is not far from the experimental value of 19.5 %. Considering the average concentrations measured before and after the experiment, the three considerations outlined above could be summarized by the following balances: Ni (−22.1 %) = O (+18.0 %) + He (+3.0 %) + neutrons (+1.1 %); Ni (−1.0 %) = Si (+0.9 %) + neutrons (+0.1 %), and Fe (−2.0 %) = Al (+1.9 %) + neutrons (+0.1 %).

9.6 Conclusions

Neutron emissions up to one order of magnitude higher than the background level were observed during the operating time of an electrolytic cell. In particular, after a time span of about 7.5 h, neutron emissions of about 3 times the background level were measured. After 9 h, it was possible to observe neutron emissions of about one order of magnitude greater than the background level. By the EDX analysis performed on the two electrodes, significant compositional variations were recorded. In general, the decrements in Pd at the first electrode seem to be almost perfectly matched by the increments in lighter elements like oxygen. As far as the second electrode is concerned, the Ni decrement is almost perfectly matched by the O and Si increments. At the same time, the Fe decrement may be considered to transform into the Al concentration after the test. The results reported in the present paper give a valid confirmation to the previous results obtained using Ni-Fe and Co-Cr electrodes, see the companion paper. Also in this case, the chemical variations and the energy emissions may be accounted for direct and indirect evidence of piezonuclear fission reactions correlated to micro-crack formation and propagation due to hydrogen embrittlement. The most relevant process emerging from the experiment is the primary fission of palladium into iron and calcium. Then, secondary fissions of both the products as well as of nickel in the other electrode appear in turn producing oxygen atoms, alpha particles, and neutrons.

References

1. Borghi DC, Giori DC, Dall'Olio A (1992) Experimental evidence on the emission of neutrons from cold hydrogen plasma. In: Proceedings of the international workshop on few-body problems in low-energy physics, Alma-Ata, Kazakhstan, Unpublished Communication (1957); Comunicacao n. 25 do CENUFPE, Recife Brazil (1971), pp 147–154
2. Diebner K (1962) Fusionsprozesse mit Hilfe konvergenter Stosswellen – einige aeltere und neuere Versuche und Ueberlegungen. Kerntechnik 3:89–93
3. Kaliski S (1978) Bi-conical system of concentric explosive compression of D-T. J Tech Phys 19:283–289
4. Winterberg F (1984) Autocatalytic fusion–fission implosions. Atomenergie-Kerntechnik 44:146
5. Derjaguin BV et al (1989) Titanium fracture yields neutrons? Nature 34:492
6. Fleischmann M, Pons S, Hawkins M (1989) Electrochemically induced nuclear fusion of deuterium. J Electroanal Chem 261:301
7. Bockris JM, Lin GH, Kainthla RC, Packham NJC, Velev O (1990) Does tritium form at electrodes by nuclear reactions? The first annual conference on cold fusion. University of Utah Research Park/National Cold Fusion Institute, Salt Lake City
8. Preparata G (1991) Some theories of cold fusion: a review. Fusion Tech 20:82
9. Preparata G (1991) A new look at solid-state fractures, particle emissions and "cold" nuclear fusion. Il Nuovo Cimento 104(A):1259–1263
10. Mills RL, Kneizys P (1991) Excess heat production by the electrolysis of an aqueous potassium carbonate electrolyte and the implications for cold fusion. Fusion Technol 20:65

11. Notoya R, Enyo M (1992) Excess heat production during electrolysis of H_2O on Ni, Au, Ag and Sn electrodes in alkaline media. In: Proceedings of the 3rd international conference on cold fusion, Nagoya Japan: Universal Academy Press, Tokyo

12. Miles MH, Hollins RA, Bush BF, Lagowski JJ, Miles RE (1993) Correlation of excess power and helium production during D_2O and H_2O electrolysis using Palladium cathodes. J Electroanal Chem 346:99–117

13. Bush RT, Eagleton RD (1993) Calorimetric studies for several light water electrolytic cells with nickel fibrex cathodes and electrolytes with alkali salts of potassium, rubidium, and cesium. In: Fourth international conference on cold fusion, Lahaina. Electric power research Institute 3412 Hillview Ave Palo Alto, CA 94304, p 13

14. Fleischmann M, Pons S, Preparata G (1994) Possible theories of cold fusion. Nuovo Cimento Soc Ital Fis A 107:143

15. Szpak S, Mosier-Boss PA, Smith JJ (1994) Deuterium uptake during Pd-D codeposition. J Electroanal Chem 379:121

16. Sundaresan R, Bockris JOM (1994) Anomalous reactions during arcing between carbon rods in water. Fusion Technol 26:261

17. Arata Y, Zhang Y (1995) Achievement of solid-state plasma fusion ("cold-fusion"). Proc Jpn Acad 71(B):304–309

18. Ohmori T, Mizuno T, Enyo M (1996) Isotopic distributions of heavy metal elements produced during the light water electrolysis on gold electrodes. J New Energy 1(3):90–99

19. Monti RA (1996) Low energy nuclear reactions: experimental evidence for the alpha extended model of the atom. J New Energy 1(3):131

20. Monti RA (1998) Nuclear transmutation processes of Lead, Silver, Thorium, Uranium. The seventh international conference on cold fusion. ENECO, Vancouver/Salt Lake City

21. Ohmori T, Mizuno T (1998) Strong excess energy evolution, new element production, and electromagnetic wave and/or neutron emission in light water electrolysis with a Tungsten cathode. Infinite Energy 20:14–17

22. Mizuno T (1998) Nuclear transmutation: the reality of cold fusion. Infinite Energy Press, New Hampshire

23. Little SR, Puthoff HE, Little ME (1998) Search for excess heat from a Pt electrode discharge in K_2CO_3-H_2O and K_2CO_3-D_2O electrolytes. Infinite Energy 5:34

24. Ohmori T, Mizuno T (2000) Nuclear transmutation reaction caused by light water electrolysis on tungsten cathode under incandescent conditions. J New Energy 4(4):66–78

25. Ransford HE (1999) Non-stellar nucleosynthesis: transition metal production by DC plasma-discharge electrolysis using carbon electrodes in a non-metallic cell. Infinite Energy 4 (23):16–22

26. Storms E (2000) Excess power production from platinum cathodes using the Pons-Fleischmann effect. In: Eighth international conference on cold fusion, Lerici (La Spezia), Italian Physical Society, Bologna, pp 55–61

27. Storms E (2007) Science of low energy nuclear reaction: a comprehensive compilation of evidence and explanations about cold fusion. World Scientific Publishing, Singapore

28. Mizuno T et al (2000) Production of heat during plasma electrolysis. Jpn J Appl Phys A 39:6055

29. Warner J, Dash J, Frantz S (2002) Electrolysis of D_2O with titanium cathodes: enhancement of excess heat and further evidence of possible transmutation. The ninth international conference on cold fusion. Tsinghua University, Beijing

30. Fujii MF et al (2002) Neutron emission from fracture of piezoelectric materials in deuterium atmosphere. Jpn J Appl Phys 41:2115–2119

31. Mosier-Boss PA et al (2007) Use of CR-39 in Pd/D co-deposition experiments. Eur Phys J Appl Phys 40:293–303

32. Swartz M (2008) Three physical regions of anomalous activity in deuterated palladium. Infinite Energy 14:19–31

33. Mosier-Boss PA et al (2010) Comparison of Pd/D co-deposition and DT neutron generated triple tracks observed in CR-39 detectors. Eur Phys J Appl Phys 51(2):20901–20911

34. Kanarev M, Mizuno T (2002) Cold fusion by plasma electrolysis of water. J Theoretics 5:1–7
35. Cardone F, Cherubini G, Petrucci A (2009) Piezonuclear neutrons. Phys Lett A 373:862–866
36. Carpinteri A, Cardone F, Lacidogna G (2009) Piezonuclear neutrons from brittle fracture early results of mechanical compression tests. Strain 45:332–339, Atti dell'Accademia delle Scienze di Torino 33: 27–42
37. Cardone F, Carpinteri A, Lacidogna G (2009) Piezonuclear neutrons from fracturing of inert solids. Phys Lett A 373:4158–4163
38. Carpinteri A, Cardone F, Lacidogna G (2010) Energy emissions from failure phenomena mechanical, electromagnetic, nuclear. Exp Mech 50:1235–1243
39. Carpinteri A, Lacidogna G, Manuello A, Borla O (2012) Piezonuclear fission reactions evidences from microchemical analysis, neutron emission, and geological transformation Rock Mech Rock Eng 45:445–459
40. Carpinteri A, Lacidogna G, Manuello A, Borla O (2013) Piezonuclear fission reactions from earthquakes and brittle rocks failure: evidence of neutron emission and nonradioactive product elements. Exp Mech 53:345–365
41. Milne I, Ritchie RO, Karihaloo B (2003) Comprehensive structural integrity: fracture of materials from nano to macro, vol 6. Elsevier, Amsterdam, pp 2, 31–33
42. Liebowitz H (1971) Fracture an advanced treatise. Academic Press, New York/San Francisco/London
43. Carpinteri A, Borla O, Goi A, Manuello A, Veneziano D (2014) Mechanical conjectures explaining cold nuclear fusion. Adv Opt Methods in Exp Mech 3:353–367. In: Conference proceedings of the society for experimental mechanics lombard, Illinois, 2013, Paper No 481
44. Veneziano D, Borla O, Goi A, Manuello A, Carpinteri A (2013) Mechanical conjectures based on hydrogen embrittlement explaining cold nuclear fusion. In: Proceedings of the 21° Congresso Nazionale di Meccanica Teorica ed Applicata (AIMETA), Torino, 2013, CD-ROM
45. Carpinteri A, Borla O, Goi A, Manuello A, Veneziano D (2016) Cold nuclear fusion explained by hydrogen embrittlement and piezonuclear fissions of the metallic electrodes: part I: Ni-Fe and Co-Cr electrodes. In: Carpinteri A et al (eds) Acoustic, electromagnetic, neutron emissions from fracture and earthquakes. Springer, New York

Chapter 10
Piezonuclear Neutron Emissions from Earthquakes and Volcanic Eruptions

Oscar Borla, Giuseppe Lacidogna, and Alberto Carpinteri

Abstract Recent neutron emission detections have led to consider also the Earth's crust, in addition to cosmic rays, as a relevant source of neutron flux variations. Neutron emissions measured in seismic areas of the Pamir region (4200 m a.s.l.) exceeded the usual neutron background level up to three orders of magnitude in correspondence to seismic activity and rather appreciable earthquakes of a magnitude of approximately the 4th degree in the Richter scale. The authors present an additional analysis with respect to those carried out by other research groups. Their studies start from recent data acquired at the "Testa Grigia" Laboratory of Plateau Rosa, Cervinia, during an experimental campaign on the evaluation of neutron radiation from cosmic rays. Even more recent data refer to dedicated experimental trials carried out in Northern Italy, at the seismic district of "Val Trebbia", Bettola, Piacenza. Moreover, the authors present the results they are obtaining at a gypsum mine located in Murisengo (Alessandria), Italy. The observations reveal a strong correlation between AE/EME/NE events and the major earthquakes in the surrounding area. The assessment of the neutron radiation at an environmental level could help to make a clear distinction between the component from the Cosmic Rays and the component from the Earth's crust.

Keywords Neutron emission • Earthquake precursors • Piezonuclear reactions • Cosmic rays

10.1 Introduction

Monitoring the different forms of energy (Acoustic Emission AE, Electro-Magnetic Emission EME, and Neutron Emission NE), emitted during the failure of natural and artificial brittle materials, enables an accurate interpretation of mechanical damage and fracture. The energy emissions have been mainly measured through the signals captured by the acoustic sensors [1–5], or the electromagnetic detectors

O. Borla (✉) • G. Lacidogna • A. Carpinteri
Department of Structural, Geotechnical and Building Engineering, Politecnico di Torino, Corso Duca degli Abruzzi 24, 10129 Torino, Italy
e-mail: oscar.borla@polito.it

© Springer International Publishing Switzerland 2015
A. Carpinteri et al. (eds.), *Acoustic, Electromagnetic, Neutron Emissions from Fracture and Earthquakes*, DOI 10.1007/978-3-319-16955-2_10

[6–13]. Nowadays, the AE technique is well-known in the scientific community and applied for structural monitoring purposes. In addition, based on the analogy between AE and seismic activity, AEs associated with microcracks are monitored, whose statistical law on frequency vs. magnitude distribution satisfies the same power-law of earthquakes. The EM signals are related to brittle materials in which the fracture propagation occurs suddenly and is accompanied by abrupt stress drops in the stress-strain diagram. A number of laboratory studies revealed the existence of EM signals during fracture experiments carried out on a wide range of materials [6].

It was also observed that the EM signals detected during failure of materials are analogous to the anomalous radiation of geoelectromagnetic waves observed before major earthquakes [7], reinforcing the idea that the EM effect can be applied as a forecasting tool for seismic events. In fact, magnetic phenomena associated to earthquakes and volcanic eruptions have been studied all over the World utilising very sensitive instruments.

As regards the neutron emissions, original experimental tests were performed by Carpinteri et al. [14–18] on brittle rock specimens. Different kinds of compression test under monotonic, cyclic, and ultrasonic mechanical loading have been carried out, fully confirming the hypothesis of piezonuclear fission reactions, giving rise to neutron emissions up to three orders of magnitude higher than the background level at the time of catastrophic failure of the specimens.

Solids that break in a brittle way are subjected to a rapid release of energy involving the generation of pressure waves that travel at a characteristic speed with an order of magnitude of 10^3 metre/second. Considering the very important case of earthquakes, it is possible to observe that, as fracture at the nanoscale (10^{-9} metres) emits phonons at the frequency scale of TeraHertz (10^{12} Hertz), so fracture at the microscale (10^{-6} metres) emits phonons at the frequency scale of GigaHertz (10^9 Hertz), at the scale of millimetre emits phonons at the scale of MegaHertz (10^6 Hertz), at the scale of metre emits phonons at the scale of kiloHertz (10^3 Hertz), and eventually faults at the kilometre scale emit phonons at the scale of the simple Hertz, which is the typical and most likely frequency of seismic oscillations. The animals with sensitive hearing in the ultrasonic field (frequency > 20 kiloHertz) "feel" the earthquake up to one day in advance, when the active cracks are still below the metre scale. Ultrasounds are in fact a well-known seismic precursor. With frequencies between Mega- and GigaHertz, and therefore cracks between the micron and the millimetre scale, phonons can generate electromagnetic waves of the same frequency, which turn out to be even a more advanced seismic precursor (up to a few days before). When phonons show frequencies between Giga- and TeraHertz, and then with cracks below the micron scale, we are witnessing a phenomenon partially unexpected: phonons resonate with the crystal lattices and, through a complex cascade of events (acceleration of electrons, bremsstrahlung gamma radiation, photo-fission, etc.), may produce nuclear fission reactions [19, 20]. It can be shown experimentally how such fission reactions can emit neutrons [14–18] like in the well-known case of uranium-235 but without gamma radiation and radioactive wastes. Note that the Debye frequency, i.e., the fundamental frequency of free vibration of crystal lattices, is around the TeraHertz, and this is not a coincidence, since it is simply due to the fact that the inter-atomic distance is just around the nanometre, as indeed the minimum size of the lattice defects. As the chain reactions are sustained by thermal neutrons in a nuclear power plant, so the piezonuclear reactions are triggered by phonons that have a frequency close to the resonance frequency of the crystal lattice and an energy close to that of thermal neutrons. Neutrons therefore appear to be as the most advanced earthquake precursor (up to two weeks before) [21–25].

These phenomena have important implications also at the Earth's crust scale. Neutron emission detections by Volodichev et al. [21, 22], Kuzhevskij et al. [23, 24], and Antonova et al. [25] have led to consider also the Earth's crust, in addition to cosmic rays, as a relevant source of neutron flux variations. Citing Volodichev et al. neutron emissions measured in seismic areas of the Pamir region (4200 m a.s.l.) exceeded the usual neutron background "up to two orders of magnitude in correspondence to seismic activity and rather appreciable earthquakes, greater than or equal to the 4th degree in the Richter scale magnitude" [22]. Considering the altitude dependence of neutron radiation (Pfotzer profile [26]), values approximately ten times higher than natural background at sea level are generally detected at 5000 m altitude. Therefore, the same earthquake occurring at sea level should produce a neutron flux up to 1000 times higher than the natural background. More recent neutron emission observations have been performed before the Sumatra earthquake of December 2004 [27]. Variations in thermal neutron measures were observed in different areas (Crimea, Kamchatka) a few days before that catastrophic event.

The authors of the present paper have recently measured neutron components exceeding the usual background in correspondence to seismic activity. These studies start from recent data acquired in the "Testa Grigia" Laboratory of Plateau Rosa, Cervinia (Italy), during an experimental campaign on neutron radiation from cosmic rays [28–30]. In particular, the assessment of the neutron radiation at the environmental level could help to make a clear distinction between the component of cosmic origin (from the Cosmic Rays) and the component from the Earth's crust (Piezonuclear Reactions).

By integrating all these signals (AE, EME, and NE) – and also considering the gaseous radon emission that appears to be one of the most reliable seismic precursors – it could be possible to set up a sort of alarm system based on a regional warning network. It could combine the signals from different alarm stations to prevent the effects of seismic events and to identify the epicentre of the earthquake. Similar networks, just based on seismic accelerations, are being utilized all over the World in Mexico, Taiwan, Turkey, Romania, and Japan [31].

Moreover, piezonuclear fission reactions related to neutron emissions from active faults may be considered as the principal cause of magnesium depletion and the consequent carbon formation during seismic activity. In this way, CO_2 atmospheric level may be considered as an appreciable precursor, together with acoustic, electromagnetic, and neutron emissions, before relevant earthquakes. Recently, significant changes in the diffuse emission of carbon dioxide were recorded in a geochemical station localized at El Hierro, in the Canary Islands

[32], before the occurrence of several seismic events during the year 2004. Appreciable CO_2 emissions were observed some days before such seismic events.

10.2 Atmospheric Neutrons at High Altitude Observatories

Galactic cosmic radiation generates secondary ionizing particles in the atmosphere (Fig. 10.1a), such as neutrons, electrons, positrons, protons, muons, and photons. Usually the dose is varying in a complicated way with altitude and with geomagnetic coordinates (longitude and latitude), being larger towards the Polar Regions and lower in the vicinity of the Equator. It also depends on the solar activity, which varies according to a cycle approximately 11 years long. Besides radiation components originating from the galactic cosmic radiation, the sun may occasionally add another component in connection with solar particle events (SPE) (Fig. 10.1b).

Usually, the neutron energy distribution is influenced by the atmospheric composition: as a matter of fact, neutrons are mainly produced by the reaction of primary protons with atmospheric nuclei N (78 %) and O (21 %).

10.2.1 Atmospheric Neutron Measurements

Since 1997, dedicated experimental campaigns [28–30] have been performed at High Altitude Observatories (HMOs), in the Northern as well as Southern Hemisphere, to obtain information on the variability of atmospheric neutron spectra with solar activity. During the research activity in these laboratories, specific techniques

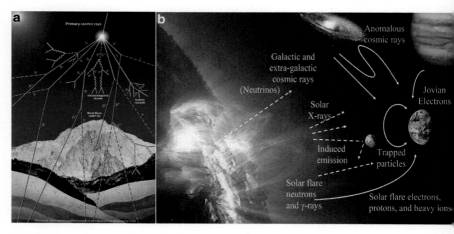

Fig. 10.1 (a) Radiation field generated by the hadronic shower in the atmosphere (*left*) (available at http://cosmicrays.le.infn.it). (b) Cosmic rays origin and composition (*right*)

for neutron spectrometry and dosimetry were set-up, suitable for being applied to different fields, such as aircrew exposure to Cosmic Rays (CRs) in high-altitude flights or in space missions. This confirms the relevance of the research activity at HMOs for the environment, space and health studies.

The experimental evaluation of neutron spectra in the wide energy range of interest and in the complex radiation field generated by the hadronic shower in the atmosphere requires a special technique. This experimental technique, based on passive neutron detectors with different threshold and energy responses, allows the reconstruction of the neutron spectra in the energy range of interest. The results of neutron spectra measurement have been obtained with passive instruments coupled with the unfolding code BUNTO, while the monitoring of the integral neutron dose has been performed by a REM (Roentgen Equivalent Man) counter.

10.2.2   Neutron Devices and Unfolding Code

The short range spectrometric system (from 10 keV to 20 MeV) is based on the passive Bubble Detector Spectrometer (BDS) (BTI, Ontario, Canada) [33]. It is constituted of polycarbonate vials filled with a tissue equivalent gel, in which tiny superheated liquid (Freon) droplets are dispersed. Neutrons interact with the gel and produce recoil charged particles, which give rise to the boiling of droplets. This leads to the formation of visible bubbles that are trapped within the gel; the number of bubbles is related to the neutron dose. Six different types of detector (with different chemical composition) are available; each of these corresponds to a different energy threshold (10, 100, 600, 1000, 2500, 10,000 keV).

The unfolding package BUNTO [34] was especially developed to process the responses of the wide and short range spectrometers. In order to get an appropriate solution from the system of Fredholm's equations, that are obtained from measurements affected by large experimental uncertainties, a special method has been introduced: it is based on the random sampling of unfolding data from a normal distribution, whose parameters (mean value and standard deviation) are the average experimental reading and the associated statistic uncertainty. The BUNTO final spectrum is the calculated mean of possible solutions of the unfolding procedure, weighted on the mean standard deviation. BUNTO fixes the maximum variation between the possible solution and the mean value within 20 %: this is assumed as "percent error" on the experimental spectrum points.

As regards the monitoring of neutron dose, the ALNOR REM counter was used. This device is able to detect the contribution of the equivalent ambient neutron dose in a very large energy range (neutron sensitivity from thermal energy to 17 MeV).

In Fig. 10.2, typical neutron spectra are shown as an example in terms of neutron fluence rate obtained at "Testa Grigia" (geographical position: 3480 m a.s.l., 45°56' N, 7°42' E) during the experimental campaigns of November 1997, March 2003 and December 2007, by using the BDS Spectrometer. The neutron spectra were measured during different solar activity periods (mean sunspot number: 10–20/11/1997: 34; 24–

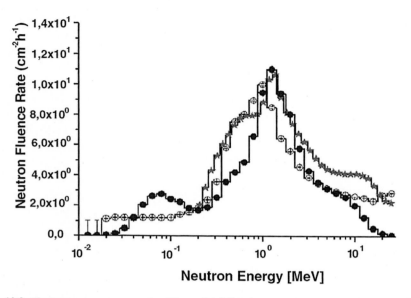

Fig. 10.2 Neutron spectra measured at "Testa Grigia" Laboratory during different solar activity periods (stars: November 1997; circles: March 2003; filled circles: December 2007), by using the BDS spectrometer (energy range 10 keV–20 MeV) [28]

31/03/2003: 88; 05–10/12/2007: 19; from the website of the National Geophysical Data Center [35]). Due to the different values of solar activity, the energy spectra show different fluence intensity and similar shape, as expected, with evidence of a main peak at about 1 MeV, the so-called evaporative contribution.

10.3   Piezonuclear Reactions: From the Laboratory to the Earth's Crust Scale

The confirmation that the environmental neutron component is linked to neutrons coming from galactic events but also from piezonuclear reactions was assessed during experimental tests conducted at Politecnico di Torino on different types of brittle rocks [14–18]. In particular, neutron emission measurements, by means of He^3 devices and neutron bubble detectors, were performed during three different kinds of compression tests: (i) under displacement control, (ii) under cyclic loading, and (iii) by ultrasonic vibration. The materials used for the tests were Luserna stones, basaltic rocks, Carrara Marble, Magnetite, and mortar enriched with iron dioxide.

During compression tests on specimens characterized by brittle material and sufficiently large size the neutron flux was found to be up to three orders of magnitude higher than the background level at the time of catastrophic failure. For specimens with more ductile behaviour, neutron emissions significantly higher

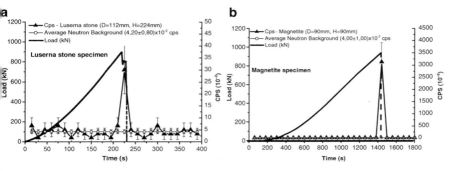

Fig. 10.3 (a) Luserna stone (*left*) and (b) Magnetite (*right*) specimens. Load vs. time diagrams, and neutron emissions count rate

than the background were also found. Neutron detection is also confirmed in compression tests under cyclic loading and during ultrasonic vibration. As an example, in Fig. 10.3a and b the load vs. time diagram, and the neutron count rate evolution for Luserna stone and magnetite specimens are shown.

Since the analyzed material contains different amounts of iron, the conjecture of Carpinteri et al. [14–18] is that piezonuclear reactions involving fission of iron into aluminum, or into magnesium and silicon, should have occurred during compression damage and failure. This hypothesis is confirmed by Energy Dispersive X-ray Spectroscopy (EDS) tests.

From the results and the experimental evidence reported in [14–18], it can be clearly seen that piezonuclear reactions are possible in inert non-radioactive solids. From the EDS results on fracture samples, the evidences of Fe and Al variations on phengite lead to the conclusion that the piezonuclear reaction:

$$\text{Fe}^{56}_{26} \rightarrow 2\,\text{Al}^{27}_{13} + 2\ \text{neutrons} \tag{10.1}$$

should have occurred [14–18, 36, 37]. Moreover, considering the evidences for the biotite content variations in Fe, Al, Si, and Mg, it is possible to conjecture that another piezonuclear reaction, in addition to (1), should have occurred during the piezonuclear tests [14–18]:

$$\text{Fe}^{56}_{26} \rightarrow \text{Mg}^{24}_{12} + \text{Si}^{28}_{14} + 4\ \text{neutrons} \tag{10.2}$$

Taking into account that granite is a common and widely present type of intrusive, Sialic, igneous rock, and that it is characterized by a high concentration in the rocks that make up the Earth's crust ($\approx 60\ \%$ of the Earth's crust), the piezonuclear fission reactions expressed above can be generalized from the laboratory to the Earth's crust scale, where mechanical phenomena of brittle fracture, due to plate collision and fault subduction, take place continuously in the most seismic areas.

10.4 Neutron Emissions from Earthquakes

As regards the observations described in [28–30], in the period from July 30 to August 3, 2008, an additional experimental campaign was conducted at the "Testa Grigia" Laboratory. These measures were performed to integrate those of December 2007. Neutron monitoring was carried out by means of the short range bubble detector spectrometer (BDS) and the REM ALNOR counter. During the data acquisition an evident increase in neutron radiation was monitored between July 31 and August 1st. This variation was detected in real time by the REM counter and later confirmed by the analysis of bubble dosimeters unfolded by BUNTO code. An increase by about six times in the neutron dose rate with respect to the average natural background was observed (Fig. 10.4a). This phenomenon was monitored for a period of about 2 h. Then the values decreased to the usual background level.

The subsequent estimation of the neutron energy spectrum (Fig. 10.4b) showed the detection of the anomalous event. In addition to the usual evaporative peak, at about 700 keV–1 MeV, a considerable high-energy neutron component of about 8 MeV was monitored. The fact that two different instruments, with different acquiring data methods, monitored simultaneously the same anomaly excludes any type of malfunction of the instrumentation. As usual, the assumptions made for the explanation of this event have firstly focused on possible effects of cosmic origin. However, from the analysis of data relating to solar and galactic events, apparently, no event of such great intensity was found. As a matter of fact, no significant sunspot activity was recorded during the data acquisition time window. As well as, during the same period, no anomalies in the cosmic ray flux were detected (Fig. 10.5) [38]. This is also demonstrated by the data acquired at the laboratory of Jungfrajoch (geographical position: 3450 m a.s.l., 46° 32′ N, 7° 59′ E), a few hundred kilometres away from the "Testa Grigia" Laboratory. In particular, in Fig. 10.5 the cosmic rays variation acquired at the laboratory of Jungfrajoch [38] in the period from July 30 to August 3, 2008, is reported. The fluctuation of few

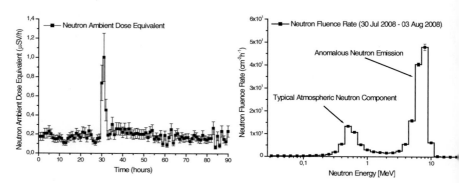

Fig. 10.4 (a) Neutron Ambient Dose Equivalent (*left*) measured by REM ALNOR counter at "Testa Grigia" Laboratory during the experimental campaign of July–August 2008. (b) Neutron spectrum (*right*) measured by using the BDS spectrometer (energy range 10 keV–20 MeV)

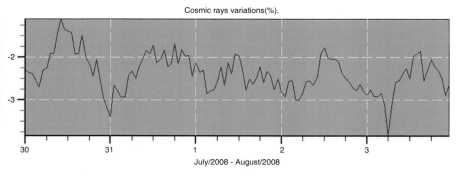

Cosmic rays variations(%).

July/2008 - August/2008

Fig. 10.5 Cosmic rays variation acquired at the laboratory of Jungfrajoch [38] in the period from July 30 to August 3, 2008

Table 10.1 Seismic activity in the surrounding area of "Testa Grigia" Laboratory in August 2008 [39]. It is also reported the distance between the earthquake epicenter and the laboratory

"Testa Grigia" Laboratory – geographical position: 3480 m a.s.l., 45°56′ N, 7°42′ E

Year	Month	Day	Latitude	Longitude	Magnitude	Distance (Km)
2008	08	10	44°18′ N	7°15′ E	2.7	199
2008	08	14	44°43′ N	7°19′ E	2.8	171
2008	08	20	44°78′ N	7°30′ E	3.0	131
2008	08	21	46°65′ N	8°47′ E	2.5	99
2008	08	21	44°86′ N	6°62′ E	2.9	145

percentages of cosmic rays flux is absolutely normal and it cannot fully explain the so marked variation in the observed environmental neutron background level shown in Fig. 10.4a and b.

On the other hand, considering the phenomenon of neutron emission before earthquakes, a searching of earthquakes occurred in the immediate vicinity of the laboratory in the weeks following the experimental campaign was carried out. A discrete seismic activity [39] was observed during the period July–August 2008 in a region a few hundred kilometres away from the laboratory (Table 10.1). In particular, about 20 days after the anomalous increase in neutron radiation, a seismic event of the 3rd degree in the Richter scale of magnitude occurred about 130 km away. This interpretation is consistent with the observations of Kuzhevskij et al. [23, 24] and it provides further experimental evidence of the correlation between neutron emissions and seismic events of appreciable intensity.

Furthermore, although some researchers impute these environmental neutron fluctuations to solar eclipses or full moon periods, which for gravitational attraction reasons can be the cause of earthquakes on the Earth's surface [21], the monitored anomalous neutron emission and the connected seismic activity occurred in a granitic geographical area, therefore strengthening the piezonuclear hypothesis, experimentally observed at the laboratory scale.

10.5 Experimental Campaign in "Val Trebbia"

From December 28, 2012, to January 6, 2013, a dedicated experimental campaign was conducted at Villanova Chiesa, Bettola, Piacenza, located in northern Italy, at the "Val Trebbia" seismic district (Fig. 10.6a) (geographical position: 410 m a.s.l., 44° 46' N, 9° 30' E). The seismic risk level of this geographical area is changed after the disastrous earthquakes that have stricken the Emilia Romagna region in Spring 2012. At the moment the area is considered a medium-high seismic zone. In Fig. 10.6b, all the seismic events observed in the year 2012 within a circular area of 100 km radius centred in Bettola are reported. A total of 218 earthquakes are detected [40], of which 194 have registered a magnitude lower than or equal to 2.5.

During the experimental trial the neutron field monitoring was carried out in "continuous mode" by means of a ^{3}He neutron radiation monitor. The AT1117M (ATOMTEX, Minsk, Republic of Belarus) neutron device is a multifunctional portable instrument with a digital readout consisting of a processing unit (PU) with an internal Geiger-Müller tube and external smart probes (BDKN-03 type). This type of device provides a high sensitivity and wide measuring ranges (neutron energy range 0.025 eV–14 MeV), with a fast response to radiation field change, ideal for environmental monitoring purposes.

Three evident peaks in neutron radiation field were monitored between December 30, 2012, and January 2, 2013. An increase by about six times in the neutron dose rate with respect to the average natural background was observed in two cases (Fig. 10.7). These phenomena were monitored for a period of at least 3 h. Then the values decreased to the usual background level.

As in the case of "Testa Grigia" Laboratory, no plausible explanation of a cosmic or galactic origin was found. As a matter of fact, only small fluctuations of few percentages in the cosmic ray flux were detected (Fig. 10.8) [38].

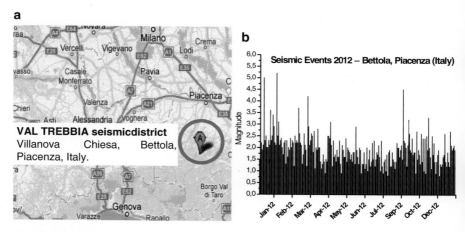

Fig. 10.6 (a) Val Trebbia seismic district. Villanova Chiesa, Bettola, Piacenza, Italy. (b) Seismic events registered in the 2012 in a circular area of 100 km centred in Bettola

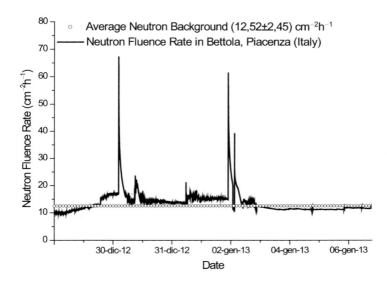

Fig. 10.7 Neutron Ambient Dose Equivalent measured by ATOMTEX radiation monitor at Villanova Chiesa, Bettola, Piacenza during the experimental campaign of December 2012–January 2013

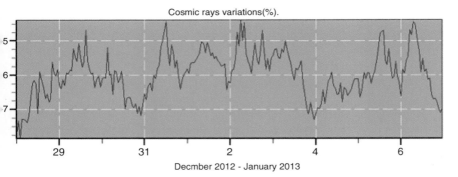

Fig. 10.8 Cosmic rays variation acquired at the laboratory of Jungfrajoch [38] in the period from December 28, 2012 to January 6, 2013

Moreover, a searching of earthquakes occurred in the immediate vicinity of the monitored area in the weeks following the experimental campaign was carried out. The usual seismic activity (average magnitude of 2.5) [40] was observed during the period from December 28, 2012, to January 6, 2013.

On the other hand, 25 days after the anomalous increase in neutron radiation an earthquake of the 5th degree in the Richter scale of magnitude was recorded (geographical position: 44°16′N 10°31′E). This event belonged to a seismic swarm occurred in the "Garfagnana" district less than 100 km from Bettola (Fig. 10.9a). In particular, In Fig. 10.9b the seismic events from December 28, 2012, to January 31, 2013, are reported.

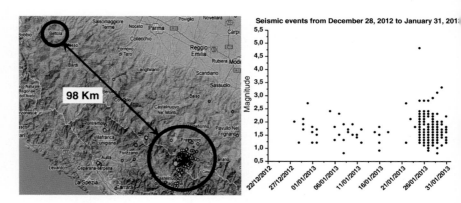

Fig. 10.9 (a) Seismic swarm occurred about 100 km away from the monitored area. (b) Seismic events from December 28, 2012 to January 31, 2013

10.6 Preliminary Results at a Gypsum Mine in Murisengo, Alessandria, Italy

Since June 24, 2013 a dedicated in-situ monitoring at the San Pietro-Prato Nuovo gypsum mine, located in Murisengo, Alessandria, Northern Italy has started and it is still in progress.

Currently, some rock pillars of the mine, located at about 100 m below the ground level, are subjected to a multi-parameter monitoring in order to assess their structural stability. The structural monitoring is principally conducted by the AE technique, but we evaluate at the same time the seismic risk of the surrounding area with the detection of the AE, EM fluctuations and the environmental neutron field.

Thanks to the position of the monitoring station (100 m under the ground level), the acoustic and electromagnetic noise of human origin is greatly reduced, as well as the neutron background is about one order of magnitude lower than on the Earth. These aspects make the mine an appropriate place for the monitoring of all the events correlated to seismic phenomena.

The preliminary results obtained during a dedicated in-situ monitoring revealed a strong correlation between AE/EME/NE events and the major earthquakes in the surrounding area. It was observed that the three different emissions anticipate the seismic event by about 1 day, 2–3 days, 1 week, respectively. They should be considered as precursors of the next earthquake rather than aftershocks of the previous one, on the basis of the different temporal distances (Table 10.2) and of the statistical signal processing. The AE hourly rate (Fig. 10.10a) and the neutron fluence rate (Fig. 10.10b) vs. the estimated local magnitude in Murisengo are reported. In both cases a power-law is obtained. By increasing the perceived magnitude, also AE and NE increase. For small earthquakes of magnitude less

Table 10.2 Acoustic and neutron emissions as seismic precursors

	PRECURSORS Time to the next earthquake (days)	AFTERSHOCKS Time after the previous earthquake (days)	Standard deviation (days)
Acoustic emissions	1.73	13.32	±0.65
Electromagnetic emissions	2.40	12.86	±0.38
Neutron emissions	6.95	8.24	±1.94

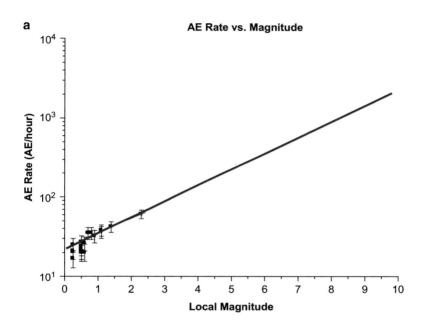

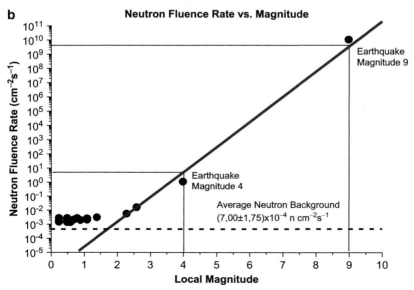

Fig. 10.10 (**a**) AE Hourly Rate vs. Local Magnitude and (**b**) Neutron Fluence rate vs. Local Magnitude

than 2.5, low acoustic emissions are monitored, whereas in the case of neutron emissions a sort of plateau at about $(1.84 \pm 0.51) \times 10^{-3}$ n cm^{-2} s^{-1} is observed. On the other hand, as conjectured in [41], considering the results described in [21–25], for higher magnitudes, a rapid increment is verified and the expected value of 10^{10} n cm^{-2} s^{-1} for seismic events of the 9th degree in the Richter scale is extrapolated.

10.7 Conclusions

Based on recent experimental data acquired at the "Testa Grigia" Laboratory of Plateau Rosa, Cervinia (Italy), further analyses – besides those already known from the literature [21–25] – were presented to confirm the hypothesis that the Earth's crust, in addition to cosmic rays, is a relevant source of neutron flux variations. This phenomenon seems to take place several days before a significant seismic activity occurs in the monitored area, or even far from the latter. In addition, preliminary results acquired at a gypsum mine and related to the evaluation of acoustic, electromagnetic and neutron emissions are reported. The experimental data emphasize the close correlation between AE/EME/NE emissions and seismic activity. In particular, it was observed that the three different emissions anticipate the seismic event by about one day, 2–3 days, one week, respectively. On the other hand, the confirmation that the environmental neutron peaks are also connected to components coming from piezonuclear fission reactions, was assessed during experimental tests on brittle rocks subjected to compression loading, conducted at the Politecnico di Torino.

In this way, a clear distinction, at the environmental level, between the neutron component of cosmic origin (from the Cosmic Rays) and that coming from the Earth's crust (piezonuclear reactions) is possible.

Taking into account the demonstrated close correlations between acoustic/electromagnetic/neutron emissions and seismic activity, it could be possible to set up a sort of alarm system combining AE, EM, and neutron detectors with radon and CO_2 concentrations, temperature variations, etc. for precurring and monitoring earthquakes.

These detectors could be applied at certain depths in the soil, along the most important faults, or very close to the most seismic areas, to prevent well in advance the effects of seismic events and to identify the epicentre of the earthquake.

Finally, regarding the possibility of prediction, the Authors are aware of the sense of caution and skepticism that has hit the international scientific community after the extensive debates arising from the catastrophic events of Fukushima in Japan and L'Aquila in Italy.

Nevertheless, the Earth's scientists now should not limit to analyze what happens after such events.

Since we are still far from the capacity of prediction of seismic events, we should consider the suggestions coming from the "innovative proposals". Obviously, these findings should not be used in substitution, but in parallel to those traditionally adopted in seismology, in order to identify the "symptoms" that can be assumed as precursors of a new seismic event.

References

1. Mogi K (1962) Study of elastic shocks caused by the fracture of heterogeneous materials and its relation to earthquake phenomena. Bull Earthquake Res Inst 40:125–173
2. Lockner DA, Byerlee JD, Kuksenko V, Ponomarev A, Sidorin A (1991) Quasi static fault growth and shear fracture energy in granite. Nature 350:39–42
3. Shcherbakov R, Turcotte DL (2003) Damage and self-similarity in fracture. Theor Appl Fract Mech 39:245–258
4. Ohtsu M (1996) The history and development of acoustic emission in concrete engineering. Mag Concr Res 48:321–330
5. Carpinteri A, Lacidogna G, Pugno N (2006) Richter's laws at the laboratory scale interpreted by acoustic emission. Mag Concr Res 58:619–625
6. Miroshnichenko M, Kuksenko V (1980) Study of electromagnetic pulses in initiation of cracks in solid dielectrics. Sov Phys Solid State 22:895–896
7. Warwick JW, Stoker C, Meyer TR (1982) Radio emission associated with rock fracture: possible application to the great Chilean earthquake of May 22, 1960. J Geophys Res 87:2851–2859
8. O'Keefe SG, Thiel DV (1995) A mechanism for the production of electromagnetic radiation during fracture of brittle materials. Phys Earth Planet Inter 89:127–135
9. Scott DF, Williams TJ, and Knoll SJ (2004) Investigation of electromagnetic Emissions in a deep underground mine. In: Proceedings of the 23rd international conference on Ground Control in Mining, Morgantown, 3–5 Aug 2004, pp 125–132
10. Frid V, Rabinovitch A, Bahat D (2003) Fracture induced electromagnetic radiation. J Phys D 36:1620–1628
11. Rabinovitch A, Frid V, Bahat D (2007) Surface oscillations. A possible source of fracture induced electromagnetic oscillations. Tectonophysics 431:15–21
12. Lacidogna G, Carpinteri A, Manuello A, Durin G, Schiavi A, Niccolini G, Agosto A (2010) Acoustic and electromagnetic emissions as precursor phenomena in failure processes. Strain 47(2):144–152
13. Carpinteri A, Lacidogna G, Manuello A, Niccolini A, Schiavi A, Agosto A (2010) Mechanical and electromagnetic emissions related to stress-induced cracks. Exp Tech 36(3):53–64
14. Carpinteri A, Cardone F, Lacidogna G (2009) Piezonuclear neutrons from brittle fracture: early results of mechanical compression tests. Strain 45:332–339
15. Cardone F, Carpinteri A, Lacidogna G (2009) Piezonuclear neutrons from fracturing of inert solids. Phys Lett A 373:4158–4163
16. Carpinteri A, Cardone F, Lacidogna G (2010) Energy emissions from failure phenomena: mechanical, electromagnetic, nuclear. Exp Mech 50:1235–1243
17. Carpinteri A, Borla O, Lacidogna G, Manuello A (2010) Neutron emissions in brittle rocks during compression tests: monotonic vs cyclic loading. Phys Mesomech 13:268–274

18. Carpinteri A, Lacidogna G, Manuello A, Borla O (2011) Energy emissions from brittle fracture: neutron measurements and geological evidences of piezonuclear reactions. Strenght Fract Complexity 7:13–31
19. Widom A, Swain J, Srivastava YN (2013) Neutron production from the fracture of piezoelectric rocks. J Phys G Nucl Part Phys 40:015006 (1–8)
20. Widom A, Swain J, Srivastava YN (2014) Photo-disintegration of the iron nucleus in fractured magnetite rocks with magnetostriction. Meccanica. doi:10.1007/s11012-014-0007-x
21. Volodichev NN, Kuzhevskij BM, Nechaev OY, Panasyuk MI, Shavrin PI (1997) Phenomenon of neutron intensity bursts during new and full moons. Cosm Res 31(2):135–143
22. Volodichev NN, Kuzhevskij BM, Nechaev OY, Panasyuk MI, Podorolsky AN, Shavrin PI (2000) Sun-Moon-Earth connections: the neutron intensity splashes and seismic activity. Astron Vestnik 34:188–190
23. Kuzhevskij M, Nechaev OY, Sigaeva EA, Zakharov VA (2003) Neutron flux variations near the Earth's crust. A possible tectonic activity detection. Nat Hazards Earth Syst Sci 3:637–645
24. Kuzhevskij M, Nechaev OY, Sigaeva EA (2003) Distribution of neutrons near the Earth's surface. Nat Hazards Earth Syst Sci 3:255–262
25. Antonova VP, Volodichev NN, Kryukov SV, Chubenko AP, Shchepetov AL (2009) Results of detecting thermal neutrons at Tien Shan high altitude station. Geomagn Aeron 49(6):761–767
26. Pfotzer G, Regener E (1935) Vertical intensity of cosmic rays by threefold coincidence in the stratosphere. Nature 136:718–719
27. Sigaeva E, Nechaev O, Panasyuk M, Bruns A, Vladimirsky B, Kuzmin Y (2006) Thermal neutrons' observations before the Sumatra earthquake. Geophys Res Abstr 8:00435
28. Zanini A, Storini M, Visca L, Durisi EAM, Fasolo F, Perosino M, Borla O, Saavedra O (2005) Neutron spectrometry at high mountain observatories. J Atmos Sol Terr Phys 67:755–762
29. Mishev A, Bouklijski A, Visca L, Borla O, Stamenov J, Zanini A (2008) Recent cosmic ray studies with lead free neutron monitor at basic environmental observatory Moussala. Sun Geosphere 3(1):26–28
30. Zanini A, Storini M, Saavedra O (2009) Cosmic rays at high mountain observatories. Adv Space Res 44(10):1160–1165
31. Allen R (2011) Seconds before the big one. Seismology, ScientificAmerican.com,54–59
32. Padron E, Melina G, Marrero R, Nolasco D, Barrancos J, Padilla G, Hernandez PA, Perez NM (2008) Changes on diffuse CO_2 emission and relation to seismic activity in and around El Hierro, Canary Islands. Pure Appl Geophys Special Issue Terrestrial Fluids, Earthquakes and Volcanoes: the Hiroshi Wakita 165(3):95–114
33. BTI (2003) Instruction manual for the Bubble Detector Spectrometer (BDS), Bubble Technology Industries, Chalk River
34. Ongaro C, Zanini A, Tommasino L (2001) Unfolding technique with passive detectors in neutron dosimetry. In: Proceedings of the workshop neutron spectrometry and dosimetry: experimental techniques and MC calculations, Stockholm, 18–20 Oct 2001, Otto Editor, pp 117–128
35. Information on National Geophysical Data Center (2012) Sunspot numbers Available at http://www.ngdc.noaa.gov/stp/solar/ssndata.html. Last accessed Apr 2012
36. Carpinteri A, Chiodoni A, Manuello A, Sandrone R (2010) Compositional and microchemical evidence of piezonuclear fission reactions in rock specimens subjected to compression tests. Strain 47(Suppl 2):282–292
37. Carpinteri A, Manuello A (2010) Geomechanical and geochemical evidence of piezonuclear fission reactions in the Earth's crust. Strain 47(Suppl 2):267–281
38. Information on Jungfraujoch Neutron Monitor (18igy) (2013) Available at http://cr0.izmiran.rssi.ru/jun1/main.htm. Last accessed Mar 2013
39. Information on National Geophysical Data Center/World Data Center (NGDC/WDC) (2012) Significant Earthquake Database, Boulder, CO, USA. Available at http://www.ngdc.noaa.gov/nndc/struts/form?t=101650&s=1&d=1. Last accessed Apr 2012

40. ISIDe Working Group (INGV, 2010). Italian Seismological Instrumental and parametric database: http://iside.rm.ingv.it. Last accessed Mar 2013
41. Carpinteri A (2016) TeraHertz phonons and piezonuclear reactions from nano-scale mechanical instabilities. In: Carpinteri A, Lacidogna G, Manuello A (eds) Acoustic, electromagnetic, neutron emissions from fracture and earthquakes. Springer, Cham

Chapter 11
Is the Shroud of Turin in Relation to the Old Jerusalem Historical Earthquake?

Alberto Carpinteri, Giuseppe Lacidogna, and Oscar Borla

Abstract Phillips and Hedges suggested, in the scientific magazine Nature (1989), that neutron radiation could be liable of a wrong radiocarbon dating, while proton radiation could be responsible of the Shroud body image formation. On the other hand, no plausible physical reason has been proposed so far to explain the radiation source origin, and its effects on the linen fibres. However, some recent studies, carried out by the first author and his Team at the Laboratory of Fracture Mechanics of the Politecnico di Torino, found that it is possible to generate neutron·emissions from very brittle rock specimens in compression through piezonuclear fission reactions. Analogously, neutron flux increments, in correspondence to seismic activity, should be a result of the same reactions. A group of Russian scientists measured a neutron flux exceeding the background level by three orders of magnitude in correspondence to rather appreciable earthquakes (4th degree in Richter Scale). The authors consider the possibility that neutron emissions by earthquakes could have induced the image formation on Shroud linen fibres, trough thermal neutron capture by Nitrogen nuclei, and provided a wrong radiocarbon dating due to an increment in C_6^{14} content. Let us consider that, although the calculated integral flux of 10^{13} neutrons per square centimetre is 10 times greater than the cancer therapy dose, nevertheless it is 100 times smaller than the lethal dose.

Keywords Shroud of turin • Neutron emission • Rocks crushing failure • Earthquake

11.1 Introduction

After the first photographs of the Shroud, taken by Mr. Secondo Pia during the Exposition of 1898 in Turin [1], a widespread interest has been generated among scientists and curious to explain the image formation and to evaluate its dating. First results of radiocarbon analyses were published in 1988. They showed that the

A. Carpinteri (✉) • G. Lacidogna • O. Borla
Department of Structural, Geotechnical and Building Engineering, Politecnico di Torino,
Corso Duca degli Abruzzi 24, 10129 Torino, Italy
e-mail: alberto.carpinteri@polito.it

© Springer International Publishing Switzerland 2015
A. Carpinteri et al. (eds.), *Acoustic, Electromagnetic, Neutron Emissions from Fracture and Earthquakes*, DOI 10.1007/978-3-319-16955-2_11

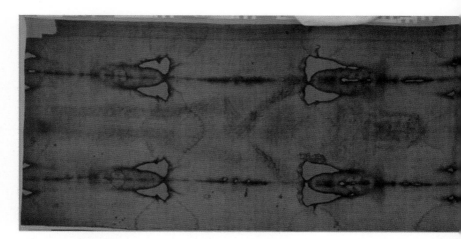

Fig. 11.1 Semitransparent front copy of the Turin Shroud (see Fanti, G., Basso, R. & Bianchini, G., Turin Shroud: Compatibility between a digitized body image and a computerized anthropomorphus manikin. Journal of Imaging and Technology 54(5), 050503 (1–8) (2010))

Shroud is at most 728 years old [2]. Later, some researchers have suggested that neutron radiation is liable of a wrong radiocarbon dating of the linen [3–5]. However, no plausible physical reason has been proposed so far to explain the radiation source origin.

Different documents in the literature attest the occurrence of disastrous earthquakes in the "Old Jerusalem" of 33 A.D., during the Christ's death [6–11]. On the other hand, recent neutron emission detections have led to consider the Earth's crust as a relevant source of neutron flux variations. Russian researchers measured neutron fluxes exceeding the background by three orders of magnitude in correspondence to seismic activity and rather appreciable earthquakes [12–17].

In this work, the authors consider that neutron emissions by earthquake –as for the conventional gadolinium-like neutron imaging technique– could have induced the image formation on Shroud linen fibres through thermal neutron capture by nitrogen nuclei, and provided a wrong radiocarbon dating due to an increment in the content of carbon-14. Moreover, some recent studies, carried out by the authors at the Laboratory of Fracture Mechanics of the Politecnico di Torino, found that it is possible to generate neutron emissions from very brittle fracture of rock specimens in compression through piezonuclear phenomena. Neutron flux increase, in correspondence to seismic activity, should be a consequence of the same phenomena [18–20].

Starting from the first photographs of the Shroud, which highlighted a figure of a human body undraped with hands crossed (Fig. 11.1), a large debate on the cause that may have produced such an image has been conducted in the scientific community. The image seems to be formed with lights and shades reversed in a sort of negative photography. Vignon [1] asserts that the image was produced by radiographic action from the body which, according to ancient texts, was wrapped in a shroud impregnated with a mixture of oil and aloes. Other authors, instead, disapprove the observations of Vignon. In particular Waterhouse [21] affirms that, if a body were wrapped in a linen cloth, under the conditions stated in the Gospels, it would be impossible for such a detailed impression to be produced in the manner suggested by Vignon.

Further studies have focused on the Shroud dating, especially since 1986 [22], when the Roman Catholic Church declared that pieces of the Shroud of Turin had been sent to seven laboratories around the world, later reduced to only three [23], for radiocarbon dating. In 1988 Dickman [2] declares that, after weeks of rumours and speculation, the official carbon dating results for the Turin Shroud were released in Zurich. The results, also published in [24], provide evidence that the linen of the Shroud of Turin is medieval, dated between 1260 and 1390. Very recently, an exhaustive study on the statistical aspects of radiocarbon dating due to the heterogeneity caused by the division of the samples into subsamples, has been also published [25].

Phillips in the paper "Shroud irradiated with neutrons?" [3] supposes that the Shroud may have been irradiated with neutrons which would have changed some of the carbon nuclei to different isotopes by neutron capture. In particular, Phillips assumes that some C_6^{14} nuclei could have generated from C_6^{13}, and that an integrated flux of 2×10^{16} thermal neutrons cm^{-2} could have produced an apparent carbon-dated age of just 670 years. However, in the reply to the same paper, Hedges [4] asserts that the integrated flux proposed by Phillips [3] is excessively high and that «including the neutron capture by nitrogen in the cloth, an integrated thermal neutron flux of 2×10^{13} would be appropriate» for the apparent radiocarbon dating of the Shroud.

Also Rinaudo [26] evaluates that simultaneous fluxes of protons and neutrons could explain at the same time the imprint on the cloth (by protons) and the thirteen-century slip in time of the C_6^{14} nuclei (by neutrons).

Fanti [27] confirms the hypothesis of Rinaudo stating that, at the same time of the neutron and proton emissions, also an electron emission could have generated the body image formation. Other Authors have recently introduced the hypothesis of radon emissions as a possible trigger of surface electrostatic discharges (ESD) and then of the image impression [28].

11.2   Earthquakes and Neutron Emissions from the Earth's Crust

Scientific data of the historical earthquake occurred around the year 33 A.D. in the Jerusalem area are mentioned in the "Significant Earthquake Database" of the American Scientific Agency NOAA (National Oceanic and Atmospheric Administration) [29]. This database contains information on destructive earthquakes from 2150 B.C. to present days. The "Old Jerusalem" earthquake is classified as an average devastating seismic event, that has also destroyed the City of Nisaea, the port of Megara, located at west of the Isthmus of Corinth [6]. It also would have involved a total cost for the reconstruction that, if the current dollar amount of damages were listed, it would be between 1.0 and 5.0 million dollars.

In addition, if we assign the image imprinted on the Shroud to the Man who died during the Passover of 33 A.D., there are at least three documents in the literature attesting the occurrence of disastrous earthquakes during and after that event.

Within his chronicle in Greek language, a historian named Thallos, probably used to lived in Rome in the middle of the first century, has left mention of events occurred on the Christ's death day: the darkening of the sky and the happening of an earthquake [7, 8]. The work of Thallos has been lost, but the quotation of the passage about Jesus had been inserted in the Chronographia of Sextus Julius Africanus, a Christian Palestinian author who died in Nicopolis around the 240 A.D.: 'The most dreadful darkness fell over the whole world, the rocks were torn apart by an earthquake and much of Judaea and the rest of the land was torn down. Thallos calls this darkness an eclipse of the sun in the third book of his Histories, without reason it seems to me. For. ...how are we to believe that an eclipse happened when the moon was diametrically opposite the sun?'.

Thallos, due to the quotation of Julius Africanus, is generally considered by historians as a witness to the early date of the gospel story of the "darkness" at the death of Christ: see Mark 15: 33; Luke 23: 44 and Mattew 27: 45 [9]. However, the interesting fact is that Julius Africanus criticizes Thallos, saying impossible that there was an eclipse on the day of Passover, which occurs in the full moon period, but he does not dispute that on the same day there was an earthquake.

On the other and, Matthew wrote that there was a strong earthquake at the moment of Christ's death: 'When the centurion and those who were with him, keeping watch over Jesus, saw the earthquake and what took place, they were filled with awe and said, "Truly this was the Son of God!"' (Matthew 27: 54) [9]. He wrote that there was another even stronger earthquake at the time of the resurrection: 'And behold, there was a great earthquake, for an angel of the Lord descended from heaven and came and rolled back the stone and sat on it. His appearance was like lightning, and his clothing white as snow. And for fear of him the guards trembled and became like dead men' (Matthew 28:2–4) [9].

There is also the narrative of Joseph of Arimathea: 'And, behold, after He had said this, Jesus gave up the ghost, on the day of the preparation, at the ninth hour. And there was darkness over all the earth; and from a great earthquake that happened, the sanctuary fell down, and the wing of the temple' (The Narrative of Joseph, Chap. 3, The good robber, 5) [10].

That event is also mentioned by Dante Alighieri, XXI Canto, Inferno, as the most violent earthquake that had ever shaken the Earth: 'Poi disse a noi: "Più oltre andar per questo / iscoglio non si può, però che giace / tutto spezzato al fondo l'arco sesto. / E se l'andare avante pur vi piace, / andatevene su per questa grotta; / presso è un altro scoglio che via face. / Ier, più oltre cinqu'ore che quest'otta, / mille dugento con sessanta sei / anni compié che qui la via fu rotta"' (Inferno, XXI Canto:106–114) [11]. Since most scholars believe that the journey of Dante began on the anniversary of the Christ's death, during the Jubilee of 1300, the chronology goes back to 33 A.D., on the Friday when, according to tradition, Christ was put to death. Therefore, it was the earthquake after the Christ's death to cause disasters and crashes, including the Sanctuary of Jerusalem, and the wing of the Solomon's Temple [10].

Nevertheless, the results from historical studies have value for Earth scientists only when the information is converted into data representing epicentral location and magnitude of the events.

Modern scholars say that Jerusalem is situated relatively close to the active Dead Sea Fault zone. They accept the occurrence of the Resurrection earthquake, to which they assign the severity of a catastrophic event, characterized by a local magnitude ML = 8.2, as well as of another earthquake that took place in Bithynia, during the same period, that would have had even a greater magnitude [12].[1]

Based on a detailed analysis of paleoearthquakes along the major active faults in the Earth's crust, some studies give evidence of their spatial and temporal distributions, as well as of their regional recurrent behaviour [30]. From these studies, it can be argued that a hypothetical earthquake of the 11th degree in the Richter scale magnitude may have a recurrence time of about 1,000 years, as well as of about 100 years one of the 10th degree, and of about 10 years one of the 9th degree.

In the active faults of the Mediterranean basin and of the Middle East region, about 1 % of the earthquake recorded during long periods over the entire surface of the Earth take place, and their historical maximum intensity should be close to the 9th degree in the Richter scale [12]. In this case, an earthquake of the 9th degree may have a recurrence time of about 1,000 years, as well as of about 100 years one of the 8th degree, and so on.

This last statistical remark would give further scientific value, as well as historical and archaeological importance, to the hypothesis that, in the "Old Jerusalem", there was a strong earthquake very close to the 9th degree in the Richter scale.

Recent neutron emission detections by Volodichev et al. [13], Kuzhevskij et al. [14, 15], and Antonova et al. [16] have led to consider the Earth's crust as a relevant source of neutron flux variations. Neutron emissions measured in seismic areas of the Pamir region (4200 m asl) exceeded the usual neutron background "up to two orders of magnitude in correspondence to seismic activity and rather appreciable earthquakes, greater than or equal to the 4th degree in the Richter scale magnitude"[13].

On the other hand, it is important to note that the flux of atmospheric neutrons increases linearly starting from the top of the atmosphere up to a maximum value corresponding to an altitude of about 20 km. From this altitude it decreases exponentially up to the sea level, where is negligible (Pfotzer profile [31]). Considering the altitude dependence of neutron radiation, values about 10 times higher than the natural background at sea level are generally detected at 5,000 m altitude. Therefore, the same earthquake occurring at sea level should produce a neutron flux up to 1000 times higher than the natural background. More recent neutron emission observations have been performed before the Sumatra earthquake of December 2004 [17]. Variations in thermal neutron measures were observed in different areas (Crimea, Kamchatka) a few days before that earthquake.

---

[1] From a geophysical and stratigraphic point of view it can be understood that the "Old Jerusalem" earthquake could have occurred between 26 and 36 A.D. [32].

11.3 Neutron Radiography and Imaging on Linen Fibres

Considering the possibility that neutron flux could have induced appreciable effects on linen fibres of the Shroud, the authors have developed some hypotheses based on piezonuclear reactions [33]. In the following, it is briefly described the process of image formation induced by neutron radiation, occurred during the earthquake in the "Old Jerusalem" of 33 A.D., and the thirteen-century slip of time effects that could be produced on linen cloths.

Neutron radiography is an imaging technique that utilizes the transmission of neutron radiation to obtain a static picture of a given object [34]. The object under examination is placed in the path of the incident radiation. The transmitted and scattered neutrons "bring the visual information" of the object that is recorded by an appropriate imaging system.

Thanks to the high thermal neutron cross section of the gadolinium nucleus (~254,000 barn), the most important detection reaction used in neutron imaging is:

$$Gd_{64}^{157} + n_0^1 \rightarrow Gd_{64}^{158} + gamma + conversion\ electrons\ (8.5\,MeV)$$

in which the converter material (gadolinium) captures neutrons and emits secondary charged particles that reproduce the irradiated object on neutron imaging plates (NIP). Usually, a thermal neutron flux of 10^5 neutrons $cm^{-2}\,s^{-1}$ is employed, with an irradiation time of few minutes for a total integrated flux of about 10^8 neutrons cm^{-2}, with typical NIP enriched for more than 20 % in weight of Gd_2O_3 [35].

Neutron imaging is a technique different, although complementary, to X-ray radiography. Whereas the X-rays are more sensitive to materials with rather high density, one of the advantages of neutron radiation is its ability to affect preferentially elements with low atomic numbers such as hydrogen, nitrogen, etc. [34]. For this reason, neutron rays give sharp images of biological soft tissue samples, whereas X-rays are more sensitive to tissues rich in calcium like bones.

The most important nuclear reaction of thermal neutrons on nitrogen nuclei is represented by:

$$N_7^{14} + n_0^1 \rightarrow C_6^{14} + H_1^1$$

that is liable of radiocarbon formation also in the atmosphere.

For the chemical composition of linen fibres, a typical nitrogen concentration of 1000 p.p.m. could be supposed [4]. The nitrogen thermal neutron capture cross section is of about 1.83 barn. By comparison, Gd_{64}^{157} cross section is 10^5 times greater than the nitrogen one. Neglecting the different concentrations in gadolinium and nitrogen, which mainly affect the image resolution, and taking into account only their cross sections, the thermal neutron flux necessary to nitrogen nuclei neutron imaging should be approximately equal to or higher than 10^{10} neutrons $cm^{-2}\,s^{-1}$.

The hypothetical reaction induced by neutrons on nitrogen nuclei might have contributed on image formation also by means of proton radiation (as assumed by Rinaudo [26]), triggering chemical combustion reactions or oxidation processes on linen fibres. Usually, image formation from a neutron beam can be accomplished in a variety of ways by using a suitable conversion screen. Thus, through an etching process with a chemical reagent (like KOH or NaOH) and under appropriate lighting, the image will become visible. Similarly, in the case of linen, neutrons could have interacted with nitrogen nuclei, and the protons – produced as secondary particles – may have assumed the function of the reagent, triggering oxidation or combustion phenomena and making the image visible. Hypotheses and experimental confirmations that oxidative phenomena generated by earthquakes can provide 3D images on the linen clothes have recently been proposed by de Liso [36].

11.4   Earthquake and Neutron Effects on the Shroud Radiocarbon Dating

Taking into account the historical sources attesting the occurrence of a disastrous earthquake in 33 A.D., and assuming a hypothetical magnitude of the 9th degree in the Richter scale [12], it is possible to provide an evaluation of the consequent neutron flux.

The Richter scale is logarithmic (base 10). This means that, for each degree increasing on the Richter scale, the amplitude and the acceleration of the ground motion recorded by a seismograph increase by 10 times. From a displacement or acceleration viewpoint, the seismic event occurred in 33 A.D. may have been 10^5 times more intense than the reference event of the 4th degree. On the other hand, from the energy viewpoint, it should have been 10^{10} times more intense than the same reference event [36].

Assuming a typical environmental thermal neutron flux background of about 10^{-3} cm^{-2} s^{-1} at the sea level, in correspondence of earthquakes with a magnitude of the 4th degree, an average thermal neutron flux up to 10^0 cm^{-2} s^{-1} should be detected, that is 1,000 times higher than the natural background [16], as previously calculated following the Pfotzer profile [31].

Thus, an earthquake of the 9th degree in the Richter scale could provide a thermal neutron flux ranging around 10^{10} cm^{-2} s^{-1}, if proportionality between released energy and neutron flux holds. A similar event could have produced chemical and/or nuclear reactions, contributing both to the image formation and to the C_6^{14} increment in the linen fibres of the Shroud, if it had totally lasted for at least 15 min. In this way, an appropriate integrated thermal neutron flux of about 10^{13} neutrons cm^{-2} is obtained, as exactly assumed by Hedges [4]. Let us consider that, although the calculated integral flux of 10^{13} neutrons per square centimetre is 10 times greater than the cancer therapy dose, nevertheless it is 100 times smaller than the lethal dose.

In confirmation of this assumption, one of the most powerful earthquakes, the so called "Greatest Chile Earthquake", occurred in Valdivia on May 22, 1960, had a complicated seismogram that lasted for at least 15 min [38]. Further information about the intensity and duration of this earthquake are reported in [39].

Neutron emissions have been detected not only at the Earth's crust scale, but also in laboratory compression experiments as shown in [18–20]. The tests have been carried out by using suitable He^3 and bubble type BD thermodynamic neutron detectors. The material employed for the tests was non-radioactive Luserna stone, a metamorphic rock deriving from a granitoid protolith. Neutron emissions from this material were found to be of about one order of magnitude higher than the ordinary natural background level at the time of the catastrophic failure. For basaltic rocks, the neutron flux achieved a level even two orders of magnitude higher than the background. In addition, a theoretical explanation is also provided in recent works by Widom et al. [40, 41].

11.5 Conclusions

Recent neutron emission detections have led to consider the Earth's crust as a relevant source of neutron flux variations. Starting from these experimental evidences, the authors have considered the hypothesis that neutron emissions from a historical earthquake have led to appreciable effects on Shroud linen fibres. Considering the historical documents attesting the occurrence in the "Old Jerusalem" of a disastrous earthquake around the year 33 A.D., the authors assume that a seismic event with magnitude ranging from the 8th to the 9th degree in the Richter scale could have produced a thermal neutron flux of up to 10^{10} cm^{-2} s^{-1}. Through thermal neutron capture by nitrogen nuclei, this event may have contributed both to the image formation, and to the increment in C_6^{14} on linen fibres of the Shroud. Let us consider that, although the calculated integral flux of 10^{13} neutrons per square centimetre is 10 times greater than the cancer therapy dose, nevertheless it is 100 times smaller than the lethal dose. Mechanical and chemical confirmations that high-frequency pressure waves, generated in the Earth's crust during earthquakes, can trigger neutron emissions were obtained at the Laboratory of Fracture Mechanics of the Politecnico di Torino, testing in compression very brittle rock specimens.

Acknowledgements This chapter represents the partial and updated version of a paper published in Scientific Research and Essays Journal, vol. 7, issue 29, pages 2603–2612, year 2012, see reference [33], and is reproduced with kind permission of Academic Journals.

References

1. Vignon M (1902) Vignon's researches and the "Holy Shroud". Nature 66:13–14
2. Dickman S (1988) Shroud a good forgery. Nature 335:663
3. Phillips TJ (1989) Shroud irradiated with neutrons? Nature 337:594
4. Hedges REM (1989) Replies to: shroud irradiated with neutrons? Nature 337:594
5. Fanti G (2012) Open issues regarding the Turin Shroud. Sci Res Essays 7(29):2504–2512

6. Mallet R (1853) Catalogue of recorded earthquakes from 1606 B.C. to A.D. 1850, Part I, 1606 B.C. to 1755 A.D. Report of the 22nd meeting of the British Association for the advancement of science held at Hull, Sept 1853, John Murray, London, pp 1–176

7. Rigg H (1941) Thallus: the samaritan? Harv Theol Rev 34:111–119

8. Prigent P (1978) Thallos, Phlégon et le Testimonium Flavianum témoins de Jésus?, Paganisme, Judaïsme, Christianisme. Influences et Affrontements dans le Monde Antique, pp 329–334 (F. Bruce Paris, 1978)

9. Gospels Acts, Matthew, Mark, Luke and John. Available at http://bible.cc. Accessed May 2011

10. The Narrative of Joseph of Arimathea. Available at http://christianbookshelf.org/. Accessed May 2011

11. Dante Alighieri, La Divina Commedia, Inferno, Canto XXI

12. Ambraseys N (2005) Historical earthquakes in Jerusalem – a methodological discussion. J Seismol 9:329–340

13. Volodichev NN, Kuzhevskij BM, Nechaev OY, Panasyuk M, Podorolsky MI (1999). Lunar periodicity of the neutron radiation burst and seismic activity on the Earth. In: Proceeding of the 26th international cosmic ray conference, Salt Lake City, August 1999, pp 17–25

14. Kuzhevskij M, Nechaev OY, Sigaeva EA (2003) Distribution of neutrons near the Earth's surface. Nat Hazards Earth Syst Sci 3:255–262

15. Kuzhevskij M, Nechaev OY, Sigaeva EA, Zakharov VA (2003) Neutron flux variations near the Earth's crust. A possible tectonic activity detection. Nat Hazards Earth Syst Sci 3:637–645

16. Antonova VP, Volodichev NN, Kryukov SV, Chubenko AP, Shchepetov AL (2009) Results of detecting thermal neutrons at Tien Shan high altitude station. Geomagn Aeron 49:761–767

17. Sigaeva E, Nechaev O, Panasyuk M, Bruns A, Vladimirsky B, Kuzmin Y (2006) Thermal neutrons' observations before the Sumatra earthquake. Geophys Res Abstr 8:00435

18. Carpinteri A, Cardone F, Lacidogna G (2009) Piezonuclear neutrons from brittle fracture: early results of mechanical compression tests. Strain 45:332–339

19. Cardone F, Carpinteri A, Lacidogna G (2009) Piezonuclear neutrons from fracturing of inert solids. Phys Lett A 373:4158–4163

20. Carpinteri A, Cardone F, Lacidogna G (2010) Energy emissions from failure phenomena: mechanical, electromagnetic, nuclear. Exp Mech 50:1235–1243

21. Waterhouse J (1903) The Holy Shroud of Turin. Nature 67:317

22. Campbell P (1986) Shroud to be dated. Nature 323:482

23. Dutton D (1988) The Shroud of Turin. Nature 332:300

24. Damon PE (1989) Radiocarbon dating of the Shroud of Turin. Nature 337:611–615

25. Riani M, Atkinson AC, Fanti G, Crosilla F (2013) Regression analysis with partially labelled regressors: carbon dating of the Shroud of Turin. J Stat Comput Simul 23:551–561

26. Rinaudo JB (1998) Image formation on the Shroud of Turin explained by a protonic model affecting radiocarbon dating. III Congresso Internazionale di Studi sulla Sindone, Torino, 5–7 June 1998

27. Fanti G (2010) Can a corona discharge explain the body image of the Turin Shroud? J Imaging Sci Technol 54(2), 020508-1/10

28. Amoruso V, Lattarulo F (2012) A physicochemical interpretation of the Turin Shroud imaging. Spec Issue Sci Res Essays 7(29):2554–2569

29. National Geophysical Data Center/World Data Center (NGDC/WDC) Significant Earthquake Database, Boulder. Available at http://www.ngdc.noaa.gov/nndc/struts/form?t=101650&s=1&d=1. Accessed May 2011

30. Min W, Zhang PZ, Deng QD (2000) Primary study on regional paleoearthquake recurrence behaviour. Acta Seismol Sin 13:180–188

31. Pfotzer G, Regener E (1935) Vertical intensity of cosmic rays by threefold coincidence in the stratosphere. Nature 136:718–719

32. Williams JB et al (2012) An early first-century earthquake in the Dead Sea. Int Geol Rev 54:1219–1228

33. Carpinteri A, Lacidogna G, Manuello A, Borla O (2012) Piezonuclear neutrons from earthquakes as a hypothesis for the image formation and the radiocarbon dating of the Turin Shroud. Sci Res Essays 7(29):2603–2612
34. Anderson IS, McGreevy R, Bilheux HZ (eds) (2009) Neutron imaging and application. A reference for the imaging community series: neutron scattering applications and techniques. Springer, New York/London, pp XVI, 341
35. Cipriani F, Castagna JC, Lehmann MS, Wilkinson C (1995) A large image-plate detector for neutrons. Physica B 213–214:975–977
36. de Liso G (2010) Shroud-like experimental image formation during seismic activity. In: Proceeding of the international workshop on the scientific approach to the acheiropoietos images, ENEA, Frascati, 4–6 May 2010
37. Richter CF (1958) Elementary seismology. W. H. Freeman, San Francisco/London
38. Barrientos SE, Ward SN (1990) The 1960 Chile earthquake: inversion for slip distribution from surface deformation. Geophys J Int 103:589–598
39. Kanamori H, Cipar JJ (1974) Focal process of the great Chilean earthquake May 22, 1960. Phys Earth Planet In 9:128–136
40. Widom A, Swain J, Srivastava YN (2013) Neutron production from the fracture of piezoelectric rocks. J Phys G Nucl Part Phys 40:15006 (1–8)
41. Widom A, Swain J, Srivastava YN (2015) Photo-disintegration of the iron nucleus in fractured magnetite rocks with magnetostriction. Meccanica 50:1205–1216

Chapter 12
Evolution and Fate of Chemical Elements in the Earth's Crust, Ocean, and Atmosphere

Alberto Carpinteri and Amedeo Manuello

Abstract The Earth's composition and evolution are topics that give rise to unanswered questions. However, some of the main evidences involving geology, geophysics and climatic equilibrium of our planet seem to imply a possible common explanation. Recently, several data, coming from geochemistry and geomechanics, have emphasized how tectonic activity should have been strictly connected to the most important changes in the Earth's Crust chemical composition over the last 4.5 Billion years. At the same time, significant measurements of neutron emissions are observed at the Earth's Crust scale during and before earthquakes. On the other hand, at the laboratory scale, original experiments performed on non-radioactive rocks under mechanical compression loading, have recently shown repeatable neutron emissions in correspondence to micro- and macro-fracture. After these experiments, a considerable reduction in the iron content appears to be consistently counterbalanced by an increase in Al, Si, and Mg contents. On these bases, the hypothesis of a new kind of nuclear reactions finds confirmations and could be considered as a valid explanation for the geologic evolution of the Earth's Crust, Ocean, and Atmosphere.

Keywords Earth's crust • Neutron emission • Chemical evolution • Piezonuclear reactions

12.1 Introduction

Over the last century, most recent scientific disciplines such as cosmology, astrophysics, and geology, have tried to answer questions concerning the origin of the Earth and the Universe [1]. Such questions have now given place to interrogatives concerning the substance that composes the Universe, the heterogeneous distribution of the main elements on the Earth, and their evolution in time [1–7].

A. Carpinteri (✉) • A. Manuello
Department of Structural, Geotechnical and Building Engineering, Politecnico di Torino,
Corso Duca degli Abruzzi 24, 10129 Torino, Italy
e-mail: alberto.carpinteri@polito.it

© Springer International Publishing Switzerland 2015
A. Carpinteri et al. (eds.), *Acoustic, Electromagnetic, Neutron Emissions from Fracture and Earthquakes*, DOI 10.1007/978-3-319-16955-2_12

Significant evidences, such as relatively abrupt changes in element abundances [2, 3, 6], the Great Oxidation Event (G.O.E.) between 2.7 and 2.4 Gyr ago [8, 9], the strong iron and nickel depletions in Earth's crust and oceans [10–12], the transition from a mafic to a sialic condition in the Continental Crust [6, 13, 14], the present level of CO_2 and N_2 concentrations in the Earth's atmosphere [8, 9], and the appreciable precursory role of CO_2 and neutron emissions before relevant earthquakes [15–20], are just some of the major events pertaining to the dynamics and the evolution of chemical element abundances in Earth's crust, oceans, and atmosphere, that still remain unsolved. Recent investigations and new instruments for data analysis led to study these unexplained phenomena more deeply.

Another question that still remains unanswered concerns the non-homogeneous composition of Oceanic and Continental Crusts. The 85 % of the Earth's volcanic eruptions take place at the sea bottom in correspondence to mid ocean-ridges [21]. These submarine volcanoes generate the solid underpinnings of all the Earth's oceans (Oceanic Crust) [2, 3, 7, 22, 23]. Comparing the data presented in the literature concerning the composition of the two different types of terrestrial crust, it can be noted that the iron concentration changes from ~8 %, in the Oceanic Crust, to ~4 % in the Continental one. Analogously, nickel changes from ~0.03 % in the Oceanic Crust to ~0.01 % in the Continental one (about a threefold decrease). Vice versa, Si, Al, and Na vary from ~24 %, ~7 %, and ~1 % in the Oceanic Crust, to ~28 %, ~8 %, and ~2.9 % in the Continental Crust, respectively [2, 3, 7]. Considering that approximately 50 % of the Continental Crust has originated, over the last 3.8 Gyr, from the Oceanic Crust subduction [2, 3, 21–23], the considerable variations in the composition of the Oceanic with respect to the Continental Crust would seem to remain a mystery [21].

12.2   Heterogeneity and Evolution of the Earth's Crust

In this context, the location of Al and Fe mineral reservoirs seems to be inexplicably connected to the geological periods when different continental zones were formed [2, 23–26]. This fact would seem to suggest that our planet has undergone a continuous evolution from the most ancient geological regions, which currently reflect the continental cores that are rich in Fe reservoirs, to more recent or contemporary areas of the Earth's Crust where the concentrations of Si and Al oxides present very high mass percentages [2, 23–26]. The main iron reservoir locations (Magnetite and Hematite mines) are reported in Fig. 12.1a [24, 26]. At the same time, the concentrations of Al-oxides and andesitic formations (the Rocky Mountains and the Andes), are shown together with the most important subduction lines, tectonic plate trenches and rifts (see Fig. 12.1b) [2, 25]. The bauxite mine locations show that the largest concentrations of Al reservoirs can be found in correspondence to the most seismic areas (Fig. 12.1b). The main iron mines are instead exclusively located in the oldest and interior parts of continents (formed through the eruptive activity of the proto-Earth), in geographic areas with a reduced

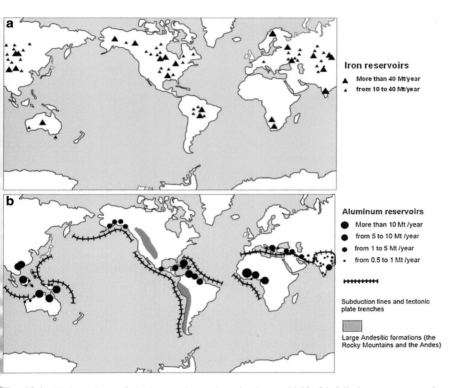

Fig. 12.1 (a) Locations of the largest iron mines in the world [2, 24–26]. Iron ore reservoirs (Magnetite and Hematite mines) are located in geographic areas with reduced seismic risks and always far from fault lines. The most abundant iron reservoirs are located in North-Central USA, Eastern Canada, North-Central Brazil, Central Australia, Ukraine, Russia, Mongolia, and North-Central China. (**b**) The largest aluminum (bauxite) reservoirs. The largest bauxite and alumina mines are located in Jamaica, Mexico, the North-Eastern littorals of Brazil, Guyana, the Gulf of Guinea, India, the Chinese littorals along the East China Sea, Greece, the South of Italy, the Philippines, New Guinea, and the Australian coast [2, 25]. The largest concentrations of Al reservoirs and the largest Andesitic formations (the Rocky Mountains and the Andes) can be found in correspondence to the most seismic areas of the Earth, subduction lines, plate tectonic trenches and rifts

seismic risk and always far from the main fault lines. What is the connection between areas rich in aluminum and poor in iron and the fault system? And why does the concentration of certain constituents seem to be connected to the geological periods in which tectonic plates were formed? These questions are still unanswered.

Many other questions concern the fact that the Earth, in contrast to the other rocky planets of our Solar System, has shown a strong compositional evolution over the last 4.57 Billion years. From 4.0 to 2.0 Gyr ago, in fact, Fe could be considered as one of the most common bio-essential elements required for the metabolic action of all living organisms [5–7, 10, 27]. Today, the deficiency of this nutrient suggests it as a limiting factor for the development of marine phytoplankton and life on Earth [6, 10, 27]. Elements such as Fe and Ni in the Earth's protocrust had higher concentrations during the Hadean (4.5-3.8 Gyr ago) and Archean (3.8−2.5 Gyr ago) periods compared to the present values [2–7, 10–14]. The Si and Al concentrations instead were lower than they are today [2–7, 14]. The estimated concentrations of Fe, Ni, Al, and Si in the Hadean and Archean Earth's protocrust and in the present Earth's Continental Crust are reported in Fig. 12.2. The data for the Hadean period (4.5−3.8 Gyr ago) are referred to the composition of Earth's protocrust, considering the assumptions made by Foing [30] and by Taylor and Mc Lennan [3]. The most abrupt changes in element concentrations shown in Fig. 12.2 appear to be intimately connected to the tectonic activity of the Earth. The vertical drops in the concentrations of Fe and Ni, as well as the vertical jumps in the concentrations of Si and Al, occurred 3.8 Gyr ago, coincide with the time that many scientists have pointed out as the beginning of tectonic activity on our planet. The subsequent step-wise transitions occurred 2.5 Gyr ago coincide with the period of the Earth's most intense tectonic activity [2, 3].

From the data reported in Fig. 12.2, a decrease of ~7 % in Fe and ~0.2 % in Ni concentrations can be observed between the Hadean period (4.5−3.8 Gyr ago) and the Archean period (3.8−2.5 Gyr ago) [2–7, 14, 28, 29]. At the same time, Al and Si concentrations increase of ~3 % and ~2 % respectively. Similarly, a global decrease of ~5 % in the concentrations of Fe (~4 %) and Ni (~1 %) and a global increase of about 3 % in the concentrations of Si (~2 %) and Al (~1 %) are shown between the Archean period (3.8−2.5 Billion years ago) and more recent times. The balances between heavier (Fe and Ni) and lighter (Si and Al) elements could be considered as perfectly satisfied, if we take into account a stepwise increase in Mg similar to that of Si over the Earth's lifetime. This Mg increase cannot be deduced from the geological data of ancient sediments but will be explained in the following considering not only the geochemical changes that involve the Earth's Crust but also other important traces that can be recognized in the evolution of the Earth's atmosphere.

In addition to Fe, Ni, Si, Al, and Mg, it is possible to consider also the other most abundant elements such as Ca, Na, K, and O, which are involved in the Earth's Crust chemical evolution. The variations in mass percentage for Mg, Ca, Na, K, and O in the Hadean and Archean Earth's protocrust and in the Earth's Continental Crust are reported in Fig. 12.3, analogously to Fig. 12.2 [2, 3, 14]. The decrease in the mass concentrations of Mg and Ca is balanced by an increase in Na, K, and O, during the Earth's lifetime. In particular, between the Hadean and the Archean eras, and between the latter and more recent times, it is possible to observe an overall decrease of ~4.7 % for Mg and ~4.0 % for Ca. This decrease in the two alkaline-earth metals (Mg and Ca) seems to be nearly perfectly balanced by the increase in the concentrations of the two alkaline metals, Na and K (which have increased by 2.7 % and 2.8 %, respectively), and by a total increase of ~3 % in O, which has varied from ~44 % to ~47 % (the latter being the present Oxygen concentration in the Earth's Crust) (Fig. 12.3). Also in this case, the greatest chemical changes in the Earth's Crust are strictly connected to the most intense tectonic activity in the planet.

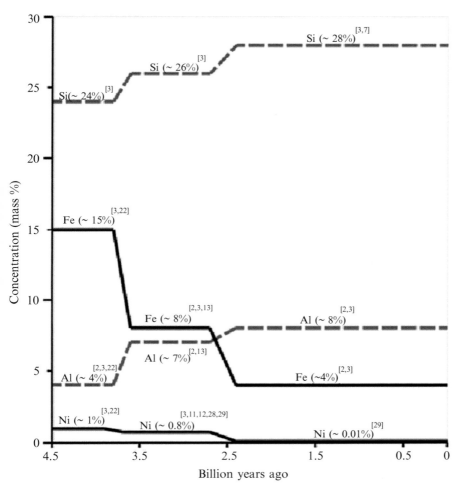

Fig. 12.2 The estimated concentrations of Fe, Ni, Al, and Si in the Hadean and Archean Earth's protocrust and in the Earth's Continental Crust are reported. The Archean Earth's protocrust (3.8−2.5 Gyr ago) had a less mafic composition (Fe ~8 %, Ni ~0.8 %, Al ~7 %, Si ~24 %) [2, 3, 10, 11, 13, 28, 29] compared to the previous period [Hadean Era, 4.5−3.8 Gyr ago) [3, 11, 12], and a less sialic composition compared to the concentrations in the Earth's Continental Crust today: Fe ~4 %, Ni ~0.01 %, Al ~8 %, Si ~28 % [2, 3, 7, 13, 14, 22, 23, 28, 29]

Plate tectonics and the connected subduction phenomena seem to involve the Earth's Crust compositional evolution and the changes in the most abundant chemical elements. These phenomena have often been ignored by orthodox views. At the same time, the macroscopical evidence emerging from Figs. 12.1, 12.2, and 12.3 cannot be explained only by the chemical element migrations or by the differentiation process that should have taken place during the Earth's formation [2, 3].

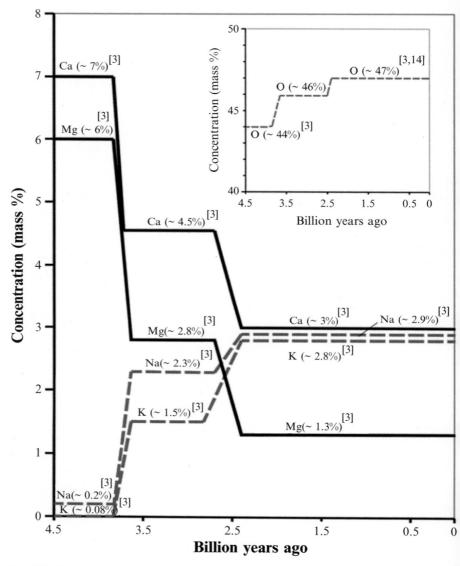

Fig. 12.3 The variations in mass percentage for Mg, Ca, Na, K, and O in the Hadean and Archean Earth's protocrust and in the Earth's Continental Crust are reported. It can be noted in particular that the overall 8.7 % decrease in alkaline-earth metals (Mg and Ca) can be nearly perfectly balanced by the Na, K, and O increases [3, 14]

According to the traditional explanation, this differentiation caused the heavy metals (iron, nickel, and other ferrous elements) to be concentrated in the core of the planet, whereas the light elements (oxygen, silicon, aluminum, calcium, potassium, sodium, etc.) were enriched in the upper layers of the Earth: the Mantle and, in particular, the Crust [2, 3]. Not everything, however, occurs simply according to density. In the authors' vision, the traditional approach cannot explain how the oceanic crust, characterized by a more basaltic composition, could generate the margins of the continents, which are composed by the lighter sialic elements. In other words, if the mechanism were only that traditionally proposed, the oceanic crust would be principally sialic as the continental one. Moreover, the traditional approach cannot explain other contradictory evidences such as that regarding the Uranium and Thorium concentrations. These two very heavy elements, should be expected to be abundant in the core, but, on the contrary, they are concentrated in the crust and the mantle of our planet.

At the same time, the temporal correlation between most important tectonic activities on Earth and most evident chemical changes in the crust, together with the nearly perfect chemical concentration balances, suggest that alternative interpretations of the Earth's crust evolution should be taken into account. Plate tectonics and high-frequency pressure waves seem to imply the chemical evolution.

This last interpretation is even more convincing if we consider recent measurements, performed by Kuzhevskij et al. [15, 16], Antonova, Volhodichev et al., and Sigaeva et al. [17–19], in correspondence to seismic activity. The results presented in these papers lead to consider also the Earth's Crust, in addition to cosmic rays, as being a relevant source of neutron flux variations. Neutron emissions exceeded the neutron background up to 1000 times in correspondence to earthquakes with a Richter magnitude equal to the 4th degree [18]. These measurements could be considered connected to the previous changes in the Earth's Crust and to mechanical phenomena of fracture, crushing, fragmentation, comminution, erosion, friction, occurring during seismic events [31–40].

12.3 From the Earth's Crust to the Laboratory: Experimental Confirmations

The macroscopic evidence of the Earth's Crust compositional evolution and the recent measurements of anomalous neutron emissions related to tectonic activity have been both confirmed by recent experiments performed at the laboratory scale. Original studies have shown a surprising although clear and repeatable evidence of anomalous neutron emissions and chemical changes obtained from the failure of non-radioactive and inert iron bearing rocks [31–40].

It had previously been shown that pressure, exerted indifferently on radioactive or inert media, can generate reproducible neutron emissions. In particular, anomalous nuclear emissions and heat generation had been verified during fracture of fissile materials [41–43] and deuterated solids [44–46], in pressurised deuterium

Table 12.1 Neutron emission level during fracture and cavitation of different materials

Material	Neutron emission
Liquids (cavitation)	
Iron chloride solution	Up to 2.5 times the background level
Solids (fracture)	
Steel	Up to 2.5 times the background level
Granite	Up to 10^1 times the background level
Basalt	Up to 10^2 times the background level
Magnetite	Up to 10^3 times the background level
Marble	Background level

gases [37], and in liquids containing radioactive deuterium and solicited by ultrasounds and cavitation [38].

The experiments recently proposed by Carpinteri et al. and by Cardone et al. [31–40] follow a different path from that of other research teams and represent the first evidence of neutron emissions from inert, stable, and non-radioactive solids under compression, as well as from non-radioactive liquids during ultrasound cavitation [47]. Successively, Carpinteri et al. have shown similar evidences varying the size of the specimens and considering catastrophic failures and scale effects on brittleness [40, 41], as well as cyclic and ultrasonic loading conditions. The maximum neutron emissions were obtained from specimens exceeding a certain volume threshold [37, 38].

Neutron emission results obtained from iron chloride liquid solutions subjected to cavitation and from non-radioactive rocks subjected to fracture, fatigue and ultrasounds are summarized in Table 12.1 [31–38, 48–50]. Liquids during cavitation were characterized by neutron emissions up to 2.5 times the background level. The same neutron emission level was obtained in the case of steel specimens subjected to compression and tension up to the final failure [37]. For granitic rocks, presenting a small amount of iron content (FeO ~1.5 %), a neutron emission up to one order of magnitude greater than the background level was detected during several tests [31–38, 48, 49]. For basalt and magnetite, where the iron concentration is much higher (FeO ~15 % and ~72.5 %, respectively), neutron emissions respectively up to 10^2 times and 10^3 times the background level were measured during fracture experiments [40]. Finally, in the case of marble no neutron emissions greater than the background level were observed during fracture experiments. This fact was firstly and erroneously explained considering the lack of iron in the marble chemical composition and the softening behavior of this material during failure [48, 49].

After these tests, Energy Dispersive X-ray Spectroscopy (EDS) analysis was performed on different sample spots of external or fracture surfaces belonging to the granite specimens used in the fracture experiments, in order to get averaged information about possible changes in their chemical composition. The measurement precision is in the order of magnitude of 0.1 %. The results for phengite and biotite (minerals with an higher Fe concentration in granite) show that the relative reduction in the iron abundance (~25 %) seems to be counterbalanced by the increment in lighter elements such as Al, Si, and Mg (Fig. 12.4) [35–37]. In

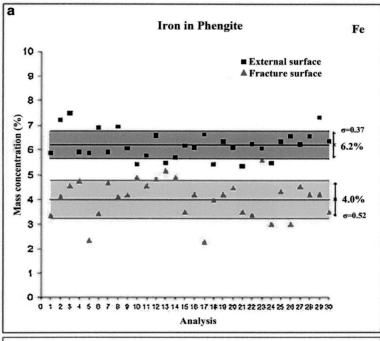

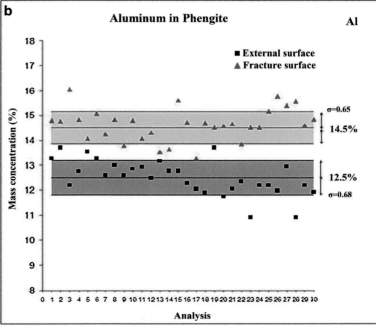

Fig. 12.4 The measurement precision is in the order of magnitude of 0.1 %. EDS results of Fe and Al concentrations in phengite analysed on samples coming from fractured specimens of granite: (**a**) Fe concentration on external surfaces (*squares*) and on fracture surfaces (*triangles*). The Fe decrease considering the two mean values of the distributions is equal to 2.2 %. (**b**) Al concentration on external surfaces (*squares*) and on fracture surfaces (*triangles*). The Al increase, considering the two mean values of the distributions, is equal to 2.0 % [34]. The evidence that the iron decrease and the Al increase are approximately equal seems to imply that reaction 1 reported in Table 12.2 occurred on fracture surfaces

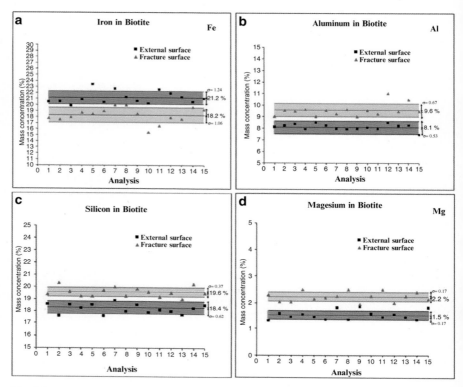

Fig. 12.5 The measurement precision is in the order of magnitude of 0.1 %. EDS results of Fe (**a**), Al (**b**), Si (**c**) and Mg (**d**) concentrations in biotite analysed on samples coming from fractured specimens of granite. Considering the analysis done on external and fracture surfaces, the iron decrease (−3.0 %) in biotite is counterbalanced by an increase in aluminium (+1.5 %), silicon (+1.2 %), and magnesium (+0.7 %) [34]. These results suggest that also reaction 12.2, reported in Table 12.2, occurred on fracture surfaces and represent a further evidence for reaction 12.1

particular, the distribution of Fe concentrations shows two different values equal to 6.2 % and 4.0 % for external and fracture surface respectively (Fig. 12.4a). Similarly, the Al concentration shows an average value changing from 12.5 % (external surface) to 14.5 % (fracture surface). The evidence that the values of the iron decrease (−2.2 %) and of the Al increase (+2.0 %) are approximately equal is really impressive.

In the Fig. 12.5a–d the results for Fe, Al, Si, and Mg concentrations of biotite crystalline phase are shown. In this case, the iron decrease is about 3.0 %, from 21.2 % (external surface) to 18.2 % (fracture surface) (Fig. 12.5a). At the same time, Al content variations show an average increase of about 1.5 % (Fig. 12.5b). In Fig. 12.5c and d it is shown that, in the case of biotite, also Si and Mg contents present considerable variations. The mass percentage of Si changes from a mean value of 18.4 % (external surface) to a mean value of 19.6 % (fracture surface), with an increase of 1.2 % (Fig. 12.5c). Similarly, the mean value of Mg concentration changes from 1.5 % (external surface) to 2.2 % (fracture surface), with an increase

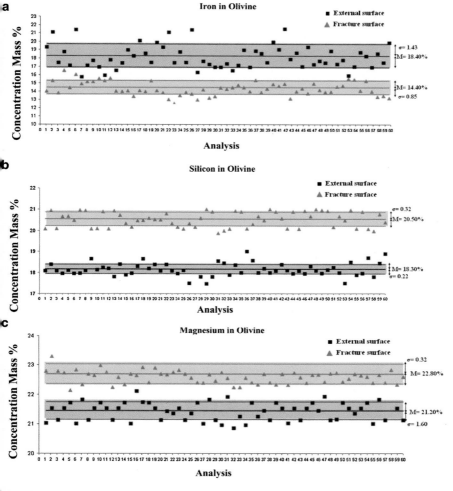

Fig. 12.6 The measurement precision is in the order of magnitude of 0.1 %. Olivine chemical changes after mechanical loading: (**a**) The Fe decrease (−4.0 %) is almost perfectly counterbalanced by an increase in Si (**b**) (+2.2 %), and Mg (**c**) (+1.6 %) [51]

of 0.7 % (Fig. 12.5d). Therefore, the iron decrease (−3.0 %) in biotite is counterbalanced by an increase in aluminum (+1.5 %), silicon (+1.2 %), and magnesium (+0.7 %) [34–37] (Table 12.2).

Similar results were observed in the case of basalt rock subjected to fatigue cycles up to the final failure. The Basalt tested in the experiments [39, 51] is dark grey with few mm-sized crystals of plagioclase, clinopyroxene and olivine. The related test was characterized by an equivalent neutron dose of about 20 times the background level. In Fig. 12.6a–c the distributions of Fe, Si and Mg concentrations are reported from 60 analysed spots on the external surface and 60 from the fracture surface. All these measurements were carried out on olivine [51]. It can be observed

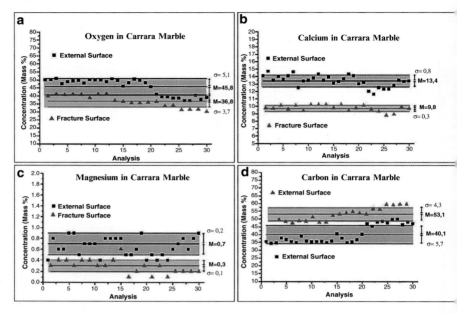

Fig. 12.7 The measurement precision is in the order of magnitude of 0.1 %. O (**a**), Ca (**b**), Mg (**c**) and C (**d**) concentrations in Carrara marble are reported for external surfaces and on fracture surfaces of samples coming from fracture experiments [52]

that the distribution of Fe content on the external surfaces shows an average value equal to 18.4 % (Fig. 12.6a). In the same graph, the distribution of Fe concentrations on the fracture samples shows a significant variation. The mean value is equal to 14.4 %, considerably lower than the mean value on the external surface (18.4 %). Similarly to Fig. 12.6a, in Fig. 12.6b the Si mass percentage concentrations are considered: the increase is approximately equal to 2.2 %. The average value of Si concentrations changes from 18.3 % on the external surface to 20.5 % on the fracture surface. In Fig. 12.6c it is shown that, in the case of basaltic olivine, also Mg content presents a considerable variation. The mass percentage concentration of Mg changes from a mean value of 21.2 % (external surface) to a mean value of 22.8 % (fracture surface) with an increase of 1.6 %. Therefore, the iron decrease (−4.0 %) in olivine is counterbalanced by an increase in silicon (+2.2 %), and magnesium (+1.6 %).

Additional results are reported for Carrara Marble specimens [52]. In this case, XPS quantitative compositional analyses were carried out in order to detect any variation in chemical composition after the brittle failure of cylinder specimens [52]. XPS survey scan (pass energy of 187.85 eV) of Carrara Marble surfaces was performed. Thirty measurements on external surface, and twenty on fracture surface were performed and analysed. In Fig. 12.7a–d, the results for the O, Ca, Mg, and C concentrations are shown. It can be observed that the distributions of O, Ca, and Mg concentrations on the external surface show average values respectively equal to 45.8 %, 13.4 %, and 0.7 %. In the same diagrams, the distributions of O,

Ca, and Mg concentrations on the fracture samples show significant variations. The mean value of the measurements performed on fracture surfaces is respectively equal to 36.8 %, 9.8 %, and 0.3 %, and they are considerably lower than the mean values on the external surface. The Carbon mass percentage increase of 13.0 % is approximately equal to the total decrease in O, Ca, and Mg. The average value of C concentrations changes from 40.1 % on the external surface to 53.1 % on the fracture surface.

12.4   Primordial Carbon Pollution and Ocean Formation: New Possible Explanations

Considering the data reported in Figs. 12.2 and 12.3 and also that granite, basalt and marble are common and widely diffused materials constituting the Earth's Continental Crust, the evidence obtained by EDS and XPS analyses, reported in Figs. 12.4, 12.5, 12.6, and 12.7, can be generalized from the laboratory to the Earth's Crust scale, where mechanical phenomena of brittle fracture, due to plate collision and subduction, take place continuously in the most seismic areas.

The most likely explanation is that, at the scale of the Earth's Crust as well as at the laboratory scale, low energy nuclear reactions (piezonuclear reactions) could take place continuously where the mechanical conditions are particularly severe. Therefore, following the evidence in the Earth's Crust (Figs. 12.1, 12.2, and 12.3) and the experimental results (Figs. 12.4, 12.5, and 12.6), the piezonuclear reactions 12.1 and 12.2 reported in Table 12.2 should have occurred during fracture phenomena [31, 32, 48, 49], without gamma ray emission or production of radioactive wastes.

Considering the present natural abundances of the major elements such as Fe, Al, Si, Mg, Na, Ni in the Continental Crust [2–7, 14, 28, 29], it is possible to conjecture that the additional nuclear reactions 12.3–12.6 reported in Table 12.2 could have taken place in correspondence to plate collision and subduction.

The large concentrations of granite minerals, such as quartz and feldspar in the Earth's Continental Crust, and to a lesser extent of magnesite, halite, and zeolite (MgO, Na_2O, Cl_2O_3), and the low concentrations of magnetite, hematite, bunsenite and cobaltite minerals (composed predominantly of Fe, Co, and Ni molecules), could be ascribed to piezonuclear reactions 12.1–12.6. The same transition, between the basaltic composition of the Oceanic Crust to the sialic composition of the Continental Crust may be explained by piezonuclear reactions 12.1–12.6.

In particular, piezonuclear reactions 12.1, 12.2, 12.5 seem to be the cause of the abrupt variations shown in Fig. 12.2. The overall 12 % decrease in the heavier elements (Fe and Ni) is perfectly balanced by the Al, Si and Mg increase (~12 %) over the last 3.8 Billion years. At 3.8 Billion years ago, in fact, we can consider the following balance: Fe (−7 %) + Ni (−0.2 %) = Al (+3 %) + Si (+2.4 %) + Mg (+1.8 %). Analogously, at 2.5 Billion years ago we have: Fe (−4 %) + Ni (− 0.8 %) = Al (+1 %) + Si (+2.4 %) + Mg (+1.4 %). The increments in Si and

Table 12.2 Piezonuclear reactions obtained by direct evidence of EDS analysis of fractured specimens and emerging from the evolution of the Continental Earth's Crust

	Earth's crust evolution
(12.1)	$Fe_{26}^{56} \rightarrow 2\,Al_{13}^{27} + 2$ neutrons
(12.2)	$Fe_{26}^{56} \rightarrow Mg_{12}^{24} + Si_{14}^{28} + 4$ neutrons
(12.3)	$Fe_{26}^{56} \rightarrow Ca_{20}^{40} + C_6^{12} + 4$ neutrons
(12.4)	$Co_{27}^{59} \rightarrow Al_{13}^{27} + Si_{14}^{28} + 4$ neutrons
(12.5)	$Ni_{28}^{59} \rightarrow 2\,Si_{14}^{28} + 3$ neutrons
(12.6)	$Ni_{28}^{59} \rightarrow Na_{11}^{23} + Cl_{17}^{35} + 1$ neutron
	Atmosphere evolution, ocean formation and origin of life
(12.7)	$Mg_{12}^{24} \rightarrow 2C_6^{12}$
(12.8)	$Mg_{12}^{24} \rightarrow Na_{11}^{23} + H_1^1$
(12.9)	$Mg_{12}^{24} \rightarrow O_8^{16} + 4H_1^1 + 4$ neutrons
(12.10)	$Ca_{20}^{40} \rightarrow 3C_6^{12} + He_2^4$
(12.11)	$Ca_{20}^{40} \rightarrow K_{19}^{39} + H_1^1$
(12.12)	$Ca_{20}^{40} \rightarrow 2O_8^{16} + 4H_1^1 + 4$ neutrons
	Greenhouse gas formation
(12.13)	$O_8^{16} \rightarrow C_6^{12} + He_2^4$
(12.14)	$Al_{13}^{27} \rightarrow C_6^{12} + N_7^{14} + 1$ neutron
(12.15)	$Si_{14}^{28} \rightarrow 2\,N_7^{14}$
(12.16)	$Si_{14}^{28} \rightarrow C_6^{12} + O_8^{16}$
(12.17)	$Si_{14}^{28} \rightarrow 2C_6^{12} + He_2^4$
(12.18)	$Si_{14}^{28} \rightarrow O_8^{16} + 2He_2^4 + 2H_1^1 + 2$ neutrons

Mg are not exactly the same due to the fact that Si is involved at the same time in reactions 12.2 and 12.5, and for the different atomic mass numbers of the two elements (see Table 12.2). However, as mentioned before, the Mg increase, due to piezonuclear reaction 12.2, cannot be revealed from geological data. The explanation is that Mg can be considered also as a starting element of further piezonuclear reactions as shown in the case of reaction 12.7. This reaction provides important explanations concerning the composition of atmosphere in the past geological eras and the present natural CO_2 emissions. The virtual increase of ~3.2 % in Mg, involved, at the same time, in reactions 12.2 and 12.7, could explain the high level of carbon in the primordial atmosphere and, consequently, the higher pressure with respect to the present one. Taking into account a density of 3.6×10^3 kg m^{-3} and a mean thickness of 60 km for the Hadean and Archean Crusts, the mass of the protocrust involved in reaction 12.7 is equal to ~3.4×10^{21} kg and implies a pressure on the Earth's surface of about 660 atm (considering the Earth's surface equal to 5.1×10^{14} m^2 like today). This pressure is very high if compared with that of the present Earth's atmosphere but can be considered a plausible value for the composition of the Earth's primordial atmosphere between 3.8 and 2.5 Billion years ago. Several authors, in fact, describe a primordial Earth's atmosphere saturated in carbon, with a pressure of several hundred atmospheres. In particular Liu [53] suggests a value of ~650 atm for the proto-atmosphere, composed principally by carbon gasses and H_2O and similar to that of Mars and Venus [53, 54]. After the intense tectonic activity that involved our planet during Adean and Archean periods, the high CO_2 concentration, principally due to piezonuclear reaction 12.7, started to decrease. This fact can be explained considering the planetary air leak of gaseous elements and molecules, such as H, He, and CO_2, that has affected the atmosphere during the Earth's lifetime [54].

Piezonuclear reaction 12.7 can also be put into correlation to the increase in seismic activity that has occurred over the last century [55]. Very recent evidence has shown CO_2 emissions in correspondence to seismic activity: significant changes in the emission of carbon dioxide were recorded in a geochemical station at El Hierro, in the Canary Islands, before the occurrence of seismic events during the year 2004. Appreciable precursory CO_2 emissions were observed to start before seismic events of relevant magnitude and to reach their maximum values some days before the earthquakes [20]. From this point of view, the evidence emerging from the XPS analyses, performed at the laboratory scale on marble, that the percentage of oxygen, calcium and magnesium totally decreases (-13.0 %) and that of C increases ($+13.0$ %), emphasizes how piezonuclear reactions 12.7, 12.10 and 12.13 may take place in calcareous rocks. In fact, no neutron emission does not imply absence of nuclear transmutations.

Reaction 12.7 is not the only one that involves Mg as a starting element. From a close examination of the data reported in Fig. 12.3, it is possible to conjecture a series of piezonuclear reactions, involving Mg, Ca, Na, K, O, C and H, that could represent the real origin of the sharp fluctuations of these chemical elements in the evolution of the Earth's Crust (see reactions 12.7–12.12 in Table 12.2). Considering piezonuclear reactions 12.8, 12.9, 12.11, 12.12, an overall decrease in alkaline-earth metals (Mg and Ca) of about 8.7 % is balanced by a nearby equal increase in Na, K, and O (see Fig. 12.3). At 3.8 Billion years ago, we have the following balance (see Fig. 12.3 and Table 12.2): Ca ($-$ 2.5 %) + Mg (-3.2 %) = K ($+1.4$ %) + Na ($+2.1$ %) + O ($+2.2$ %). At 2.5 Billion years ago, on the other hand, we have: Ca (-1.5 %) + Mg (-1.5 %) = K ($+1.3$ %) + Na ($+0.6$ %) + O ($+1.1$ %). Also in this case, the mass percentage variations in the major elements of the Earth's Crust could be perfectly explained considering this new kind of nuclear fission reactions (see Fig. 12.3).

In particular, the global decrease in Ca (~4.0 %) can be counterbalanced by an increase in K (~2.7 %), reaction 12.11, and in H_2O (~1.3 %), reaction 12.12. Considering the mass of the proto-crust equal to ~1.08×10^{23} kg, the decrease of about 1.3 % in Ca concentration corresponds to ~1.40×10^{21} kg. This value is very close to the mass of water in the oceans ~1.35×10^{21} kg (considering a global ocean surface of ~3.60×10^{14} m^2 and an average depth of 3950 m). In this way, reaction 12.12 could be considered as responsible for the formation of oceans during the Earth's life time. These last considerations seem to be even more important taking into account the recent results from Europe's Rosetta mission about the origin of water on the planet Earth. The evidences coming from the exploration of the Comet 67P returned significant difference comparing the water in comets with the water we have on our planet leading to exclude the extraterrestrial origin of the ocean [56].

Moreover, taking into account reaction 12.6, a small portion of the overall Ni decrease reported in Fig. 12.2 could be counterbalanced by the salinity level in the oceans. Considering the NaCl concentration (~3.8 %) in the sea water, the mass of dissolved sodium chloride can be quantified in 5.14×10^{19} kg. This value corresponds to about 0.05 % of the Earth's proto-crust and could be ascribed to a small part of the total Ni diminution between Archean period and present time (~0.8 %). This conjecture could be justified considering: (i) the increase in NaCl concentrations in the Earth's oceans, about twice, between 2.7 and 2.4 Gyr ago [22, 57]; (ii) the very high level of salinity in the Ionian and Aegean seas [27].

Piezonuclear reactions 12.13–12.18, involving Si, Al and O as starting elements and C, N, O, H, and He as resultants, can be considered (see Table 12.2) for greenhouse gas formation. On the other hand, piezonuclear reactions 12.1–12.12 are those that have affected the evolution of the Earth's Crust over the last 4.5 Billion years to the greatest extent.

Considering the data shown in Figs. 12.2 and 12.3, it can be noted that reactions 12.1, 12.2, 12.5, 12.8, 12.11, 12.12 should have been particularly recurrent and responsible for a variation of approximately 21 % in the chemical composition of the Earth's Crust. Piezonuclear reactions 12.7, 12.9, 12.10, 12.13–12.18 have instead played a negligible role in the compositional variations of the Earth's Crust, but are of a great relevance as far as the increase in H, He, C, N, and O concentrations is concerned in the evolution of the Earth's atmosphere [53]. Reactions 12.9, 12.12, 12.16, and 12.18 played also a significant role in phenomena such as the Great Oxidation Event (2.7–2.4 Billion years ago), with a sharp increase in atmospheric oxygen (10^5-fold increase) and the subsequent formation of oceans and origin of life, as well as a high level of N_2, in the primordial atmosphere.

12.5 Conclusions

The piezonuclear fission reactions identified by the authors can be subdivided into two different sets that are typical of our planet:

$$Fe_{26}, \; Co_{27}, \; Ni_{28} \;\; \rightarrow \;\; Mg_{12}, \; Al_{13}, \; Si_{14} \;\; \rightarrow \;\; C_6, \; N_7, \; O_8$$

These two piezonuclear jumps represent not only the temporal variations in the Earth's Crust composition over the last four Billion years, but also reflect the spatial variations in the Earth composition from the internal core to the atmosphere: core (Ni-Fe and Co alloys), mantle (Si-Ma), crust (Si-Al), and atmosphere (C, N, O).

In addition, as shown by reactions 12.1, 12.2, the fundamental starting element in the first jump is iron. This element plays also a crucial role in the stellar thermonuclear fusion process of small and medium sized stars. Fe, in fact, is the heaviest element that is synthesized by thermonuclear fusion reactions before the star collapse. This combustion product is the so-called stellar ash and is the material of which the proto-planets in the Solar System were formed and from which they may evolve, through piezonuclear fission reactions, to a sialic condition (see the case of the Earth). Similarly, the atmospheric elements (C, N, O, H, and He) can be considered as the planetary ashes. These gases are the results of the second piezonuclear jump and they have gradually escaped from Earth into space through the planetary air leak [54]. The great number of evidences reported herein cannot be considered as mere coincidences but as a multiple signature of the same phenomenon, the piezonuclear fission reactions, taking place everywhere, from the laboratory to the planetary scale, and everytime, from the origin of Earth to the present time.

Acknowledgments The authors have not been financially supported by any specific grant or agency for this research. However, they would like to anticipatedly acknowledge any form of support to further future studies in the same directions. Special thanks are due to Prof. R. Sandrone of the Environmental Engineering, Land and Infrastructures Department (Politecnico di Torino) and Dr. A. Chiodoni of the Italian Institute of Technology (IIT) for their extensive collaboration in the EDS analysis.

References

1. Kolb E (2000) Blind watchers of the sky: the people and ideas that shaped our view of the universe. Oxford University Press, Oxford
2. Favero G, Jobstraibizer P (1996) The distribution of aluminum in the Earth: from cosmogenesis to Sial evolution. Coord Chem Rev 149:367–400
3. Taylor SR, McLennan SM (2009) Planetary Crusts: their composition, origin and evolution. Cambridge University Press, Cambridge
4. Hawkesworth CJ, Kemp AIS (2006) Evolution of the continental crust. Nature 443:811–817
5. Canfiled DE (1998) A new model for Proterozoic ocean chemistry. Nature 396:450–453
6. Anbar AD (2008) Elements and evolution. Science 322:1481–1482
7. Doglioni C (2007) Interno della Terra, Treccani, Enciclopedia Scienza e Tecnica, 595–605
8. Holland HD (2006) The oxygenation of the atmosphere and oceans. Philos Trans R Soc Lond Ser B 361:903–915
9. Kump LR, Barley ME (2007) Increased subaerial volcanism and the rise of atmospheric oxygen 2.5 Billion years ago. Nature 448:1033–1036
10. Buesseler KO et al (2008) Ocean iron fertilization moving forward in a sea of uncertainty. Science 319:162
11. Saito MA (2009) Less nickel for more oxygen. Nature 458:714–715
12. Konhauser KO et al (2009) Oceanic nickel depletion and a methanogen famine before the Great Oxidation Event. Nature 458:750–754
13. Rudnick RL, Fountain DM (1995) Nature and composition of the continental crust: a lower crustal perspective. Rev Geophys 33(3):267–309
14. Yaroshevsky AA (2006) Abundances of chemical elements in the Earth's crust. Geochem Int 44(1):54–62
15. Kuzhevskij BM, Yu. Nechaev O, Sigaeva EA, Zakharov VA (2003) Neutron flux variations near the Earth's Crust. A possible tectonic activity detection. Nat Hazards Earth Syst Sci 3:637–645

16. Kuzhevskij BM, Nechaev OY, Sigaeva EA (2003) Distribution of neutrons near the Earth's surface. Nat Hazards Earth Syst Sci 3:255–262
17. Antonova VP, Volodichev NN, Kryukov SV, Chubenko AP, Shchepetov AL (2009) Results of detecting thermal neutrons at Tien Shan High Altitude Station. Geomagn Aeron 49:761–767
18. Volodichev NN, Kuzhevskij BM, Nechaev OY, Panasyuk MI, Podorolsky AN, Shavrin PI (2000) Sun-Moon-Earth connections: the neutron intensity splashes and seismic activity. Astron Vestn 34(2):188–190
19. Sigaeva E et al (2006) Thermal neutrons'observations before the Sumatra earthquake. Geophys Res Abstr 8:00435
20. Padron E et al (2008) Changes in the diffuse CO_2 emission and relation to seismic activity in and around El Hierro, Canary Islands. Pure Appl Geophys 165:95–114
21. Kelemen PB (2009) The origin of the land under the sea. Sci Am 300(2):42–47
22. Hazen RM et al (2008) Mineral evolution. Am Mineral 93:1693–1720
23. Lunine EJI (1998) Earth: evolution of a habitable world. Cambridge University Press, Cambridge/New York/Melbourne
24. Roy I, Sarkar BC, Chattopadhyay A (2001) MINFO-a prototype mineral information database for iron ore resources of India. Comput Geosci 27:357–361
25. World Mineral Resources Map. Available at http://www.mapsofworld.com/world-mineral-map.htm. Last accessed Oct 2009
26. Key Iron Deposits of the World. Available at http://www.portergeo.com.au/tours/iron2002/-iron2002depm2b.asp. Last accessed Oct 2009
27. Sigman D, Jaccard S, Haug F (2004) Polar ocean stratification in a cold climate. Nature 428:59–63
28. Egami F (1975) Minor elements and evolution. J Mol Evol 4(2):113–120
29. National Academy of Sciences (1975) Medical and biological effects of environmental pollutants: nickel. Proc. Natl Acad Sci, Washington, DC
30. Foing B (2005) Earth's childhood attic. Astrobiol Mag. http://link.springer.com/article/10.1007%2FBF01732017#
31. Carpinteri A, Lacidogna G, Manuello A, Borla O (2010) Piezonuclear transmutations in brittle rocks under mechanical loading: microchemical analysis and geological confirmations. In: Kounadis AN, Gdoutos EE (eds) Recent advances in mechanics. Springer, Chennai, pp 361–382
32. CarpinteriA, Lacidogna G, Manuello A, Borla O (2010) Piezonuclear transmutations in brittle rocks under mechanical loading: microchemical analysis and geological confirmations. In Recent advances in mechanics. Dedicated to the Late Professor P.S. Theocaris, Springer, Chennai, pp 361–382
33. Carpinteri A, Borla O, Lacidogna G, Manuello A (2010) Neutron emissions in brittle rocks during compression tests: monotic vs. cyclic loading. Phys Mesomech 13:264–274
34. Carpinteri A, Chiodoni A, Manuello A, Sandrone R (2011) Compositional and microchemical evidence of piezonuclear fission reactions in rock specimens subjected to compression tests. Strain 47:267–281
35. Carpinteri A, Lacidogna G, Manuello A, Borla O (2011) Energy emissions from brittle fracture: neutron measurements and geological evidences of piezonuclear reactions. Strength Fract Complex 7:13–31
36. Carpinteri A, Lacidogna G, Borla O, Manuello A, Niccolini G (2012) Electromagnetic and neutron emissions from brittle rocks failure: experimental evidence and geological implications. Sadhana 37:59–78
37. Carpinteri A, Lacidogna G, Manuello A, Borla O (2012) Piezonuclear fission reactions: evidences from microchemical analysis, neutron emission, and geological transformation. Rock Mech Rock Eng 45:445–459
38. Carpinteri A, Lacidogna G, Manuello A, Borla O (2013) Piezonuclear fission reactions from earthquakes and brittle rocks failure: evidence of neutron emission and nonradioactive product elements. Exp Mech 53(3):345–365
39. Carpinteri A, Manuello A (2012) An indirect evidence of piezonuclear fission reactions: geomechanical and geochemical evolution in the Earth's Crust. Phys Mesomech 15:14–23

40. Carpinteri A, Borla O, Lacidogna G, Manuello A (2012) Piezonuclear reactions produced by brittle fracture: from laboratory to planetary scale. In: Proceedings of the 19th European conference of fracture, Kazan
41. Diebner K (1962) Fusionsprozesse mit Hilfe konvergenter Stosswellen – einige aeltere und neuere Versuche und Ueberlegungen. Kerntechnik 3:89–93
42. Kaliski S (1976) Critical masses of mini-explosion in fission–fusion hybrid systems. J Tech Phys 17:99–108
43. Winterberg F (1984) Autocatalytic fusion–fission implosions. Atom-Kerntech 44:146
44. Derjaguin BV et al (1989) Titanium fracture yields neutrons? Nature 34:492
45. Preparata G (1991) A new look at solid-state fractures, particle emissions and "cold" nuclear fusion. Il Nuovo Cimento 104 A:1259–1263
46. Fujii MF et al (2002) Neutron emission from fracture of piezoelectric materials in deuterium atmosphere. Jpn J Appl Phys 41(Pt.1):2115–2119
47. Carpinteri A (1989) Cusp catastrophe interpretation of fracture instability. J Mech Phys Solids 37:567–582
48. Carpinteri A, Cardone F, Lacidogna G (2009) Piezonuclear neutrons from brittle fracture: early results of mechanical compression tests. Strain 45:332–339. Presented at the Turin Academy of Sciences on 10 Dec 2008. Proc Turin Acad Sci 33:27–42
49. Cardone F, Carpinteri A, Lacidogna G (2009) Piezonuclear neutrons from fracturing of inert solids. Phys Lett A 373:4158–4163
50. Cardone F, Cherubini G, Petrucci A (2009) Piezonuclear neutrons. Phys Lett A 373 (8–9):862–866
51. Manuello A, Sandrone R, Guastella S, Borla O, Lacidogna G, Carpinteri A (2012) Piezonuclear reactions during mechanical tests of basalt and magnetite. In: Proceedings of the 19th European conference of fracture, Kazan
52. Lacidogna G, Borla O, Carpinteri A (2012) X-ray Photoelectron Spectroscopy on fracture surfaces of Carrara marble specimens crushed in compression. In: Proceedings of the 19th European conference of fracture, Kazan
53. Liu L (2004) The inception of the oceans and CO_2-athmosphere in the early history of the Earth. Earth Planet Sci Lett 227:179–184
54. Catling CD, Zahnle KJ (2009) The planetary air leak. Sci Am 300(5):24–31
55. Aki K (1983) Strong motion seismology. In: Kanamori H, Boschi E (eds) Earthquakes: observation, theory and interpretation. North Holland Pub. Co, Amsterdam, pp 223–250
56. Altwegg K et al (2014) 67P/Churyumov-Gerasimenko, a Jupiter family comet with a high D/H ratio. Science. doi:10.1126/science.1261952
57. Knauth P (1998) Salinity history of the Earth's early ocean. Nature 395:554–555

Chapter 13
Chemical Evolution in the Earth's Mantle and Its Explanation Based on Piezonuclear Fission Reactions

Alberto Carpinteri, Amedeo Manuello, and Luca Negri

Abstract The anomalous chemical balances at the major events in the geomechanical and geochemical evolution of the Earth's crust should be considered as indirect evidences of piezonuclear fission reactions. Recent results observed at the scale of the Earth's crust and reproduced at the scale of the laboratory during quasi-static and repeated loading experiments may be extended to the different layers of the planet like the atmosphere and the bulk Earth (mantle and external core). The mantle of our planet is characterized by very high pressures and temperatures (~150 GPa and ~4000 °C) that could favour this kind of reactions. In the present paper, it is shown that the most important chemical changes in the Earth's crust evolution may be recognized also at the internal Earth's layers. Recent investigations have shown that also the mantle is characterized by significant compositional time variations. This evolution may be interpreted in the light of the same nuclear reactions recently proposed to explain the chemical changes in the Earth's continental crust and atmosphere through the entire life of our planet.

Keywords Earth's mantle • Chemical evolution • Piezonuclear fission reactions

13.1 Introduction

Recent geophysical and geochemical studies of our planet present a rich array of large-scale processes and phenomena that are not fully understood [1]. These range from the subducted slabs to the nature of the core – mantle boundary; from the mechanisms forming the present-day crust, mantle, and core to the distribution of Earth's crust elements and the uptake and recycling of volatiles throughout Earth's history [1]. It is only recently that pressures similar to those prevailing inside the Earth can be produced in the laboratory, as well as the materials can be tested by the necessary tools, in order to study the interior Earth's composition. The experiments

A. Carpinteri (✉) • A. Manuello • L. Negri
Department of Structural, Geotechnical and Building Engineering, Politecnico di Torino, Corso Duca degli Abruzzi 24, 10129 Torino, Italy
e-mail: alberto.carpinteri@polito.it

© Springer International Publishing Switzerland 2015
A. Carpinteri et al. (eds.), *Acoustic, Electromagnetic, Neutron Emissions from Fracture and Earthquakes*, DOI 10.1007/978-3-319-16955-2_13

have demonstrated that, under such extreme conditions, the physical behavior of materials can be profoundly altered, causing new and unforeseen reactions and giving rise to electromagnetic and nuclear transitions in rocks and minerals [1]. Similar results have been recently observed also in the rocks of near-surface environment. During the last few years, in fact, surprising results have been reported considering the chemical changes that characterized the Earth's crust and atmosphere evolution, together with the ocean formation, during the last 4.57 Billion years [2–9]. These evidences are related to the depletions of elements such as iron and nickel and the increment in lighter ones just in correspondence to tectonic activities in our planet [3–5]. The chemical changes observed at the scale of the Earth's crust have been interpreted by innovative experiments conducted on non-radioactive rocks subjected to different mechanical loading conditions [10–13]. These rocks are natural materials that represent the composition of the Earth's near-surface environment, such as granite, basalt, magnetite, and calcareous rocks. All these results provided evidence concerning a new kind of nuclear reactions that may take place during static or cyclic-fatigue tests [10]. The phenomena observed at the laboratory scale suggested that pressure waves of very high frequency, suitably exerted on inert and stable nuclides, generate nuclear reactions of a new type accompanied by evident and reproducible chemical changes localized on the fracture surfaces of rock fragments [10–16].

The catalysing factor producing these new kind of nuclear reactions, with so important geological and geophysical effects, is pressure. In particular, these reactions would be activated where the environment conditions (pressure and temperature) are particularly severe, and mechanical phenomena of fracture, crushing, fragmentation, comminution, erosion, friction, etc., may occur [8–14]. It is evident that the physical conditions characterising the microseismic and seismic activity of an impending earthquake may be sufficient to produce these nuclear reactions at the scale of the Earth's crust. Even more severe conditions regard the bulk of our planet [1]. The mantle, presenting huge pressures and temperatures reaching ~150 GPa and ~4000 °C, respectively, is characterized by significant compositional time variations, similar to those observed in the Earth's crust [1, 3, 17–20]. As will be reported in the present paper, the chemical evolution of the Mantle may be interpreted in the light of the same low energy nuclear reactions recently discovered and firstly proposed to explain the chemical evolution of the Earth's lithosphere.

13.2 Earth's Crust and Atmosphere Evolution

A transition from a mafic to a sialic composition of the Earth's Crust was observed by different authors [2, 3, 8, 9]. Recent studies advanced the hypothesis that the changes in Ni, Fe, Si, and Al compositions can be interpreted by the so-called piezonuclear fission reactions [8–13, 15, 16]. During the first Earth's geological period, when the planet's tectonic activity began (3.8 Gyr ago), the decrease in Fe

Table 13.1 Quantitative data for the "first piezonuclear jump" in the Earth's crust

Geological period	Decrease of Fe and increase of Al, Si and Mg				Decrease of Ni and increase in Si and NaCl		
	Fe (%)	Al (%)	Si (%)	Mg (%)	Ni (%)	Si (%)	(NaCl) (%)
Beginning of Earth's tectonic activity 3.8 Gyr ago	−7	+3	+2.2	+1.8	−0.2	+0.2	+0
Greatest tectonic activity 2.5 Gyr ago	−4	+1	+1.6	+1.4	−0.8	+0.75	+0.05
Total per element	−11	+4	+3.8	+3.2	−1	+0.95	+0.05

amounted to 7 % of the mass of the Hadean protocrust, and through the symmetric piezonuclear reaction [8–13] (see also Table 13.1):

$$Fe_{26}^{56} \rightarrow 2\ Al_{13}^{27} + 2\,neutrons \tag{13.1}$$

this was partially counterbalanced by a 3 % increase in Al. The remaining 4 % can be explained considering the asymmetric fission reaction (see also Table 13.1).

$$Fe_{26}^{56} \rightarrow Mg_{12}^{24} + Si_{14}^{28} + 4\ neutrons. \tag{13.2}$$

In evaluating the effect of this reaction, we can consider a 2.2 % increase in Si and an increase in Mg of 1.8 % (see Table 13.1). In the same period, the decrease in Ni appears to be 0.2 %. The latter reduction can be interpreted in the light of the following reaction:

$$Ni_{28}^{59} \rightarrow 2\ Si_{14}^{28} + 3\ neutrons \tag{13.3}$$

which was balanced by an identical 0.2 % increase in Si. From these reactions, the overall increase in Si for this period is thus around 2.4 % (see Table 13.1).

In the second geological period of the Earth's history, at the most intense tectonic activity (2.5 Gyr ago), Fe decreased by 4 %, and this can be balanced by a 1 % increase in Al through reaction (13.1). With regard to reaction (13.2), the remaining 3 % can be balanced by an increase in Si (+1.6 %) and in Mg (+1.4 %). In the same period, 0.8 % decrease in Ni can, by taking reaction (13.3) into account, be balanced by an increase in Si of the same percentage. Nickel can also be involved in the following piezonuclear reaction in addition to the reaction presented above:

$$Ni_{28}^{59} \rightarrow Na_{11}^{23} + Cl_{17}^{35} + 1\ neutron \tag{13.4}$$

The recent studies cited above [8–15] also illustrate the compositional changes in the Earth's crust that have taken place for Ca, Mg, K, Na, and O. In the first geological critical period (3.8 Gyr ago), the decrease in Ca amounts to 2.5 %, through the reaction:

$$Ca_{20}^{40} \rightarrow K_{19}^{39} + H_1^1 \tag{13.5}$$

Table 13.2 Quantitative data for the "second piezonuclear jump" in the Earth's crust

Geological period	Decrease of Ca and increase of K and H_2O			Decrease of Mg and increase of Na, O and H		
	Ca (%)	K (%)	H_2O (%)	Mg (%)	Na (%)	O + H (%)
Beginning of Earth's tectonic activity 3.8 Gyr ago	−2.5	+1.4	+1.1	−3.2	+2.1	+1.1
Greatest tectonic activity 2.5 Gyr ago	−1.5	+1.3	+0.2	−1.5	+0.6	+0.9
Total per element	−4.0	+2.7	+1.3	−4.7	+2.7	+2.0

This decrease is balanced by a 1.4 % increase in K. The remaining 1.1 % may be balanced by the following reaction (see also Table 13.2):

$$Ca_{20}^{40} \rightarrow 2O_8^{16} + 4H_1^1 + 4\text{neutrons} \tag{13.6}$$

This reaction may be considered to explain the formation of H_2O, which was firstly concentrated in the primordial atmosphere and then condensed to form the oceans (see Table 13.2). In the same critical period, there was a 3.2 % reduction in Mg:

$$Mg_{12}^{24} \rightarrow Na_{11}^{23} + H_1^1 \tag{13.7}$$

which can be balanced by a 2.1 % increase in Na and by a total increase in O and H of 1.1 % (see Table 13.2). In the second critical geological period (2.5 Gyr ago), the decrease in Ca is 1.5 %, which, considering reaction (13.5), is balanced by a 1.3 % increase in K. Taking reaction (13.6) into account, the remaining 0.2 % is balanced by the formation of H_2O (see Table 13.2). The 1.5 % decrease in Mg for the same period, see reaction (13.7), can be balanced by a 0.6 % increase in Na and by a total increase in O and H of 0.9 % (see Table 13.2). In both critical periods of compositional step-wise transition (3.8 and 2.5 Gyr ago), the reduction in Ca that is not balanced by an increase in K can be balanced by the formation of H_2O molecules, see reaction (13.6) (see Table 13.2).

13.2.1 Atmosphere Evolution and Climate Change

Recent studies have demonstrated that the Earth's atmosphere, as well as its crust has undergone significant changes in its major constituent elements over the last 4.5 billion years [17–26].

The findings presented here stem from studies of the evolution of the individual gases in the early atmosphere and considerations regarding the variations in the atmospheric pressure exerted on Earth's primordial crust. These analyses make it possible to formulate hypotheses about both the constitution of the atmosphere and the early tectonic phenomena that took place during the Hadean (4.5–3.8 billion years ago) and Archean (3.8–2.5 billion years ago) periods.

In the early stages of Earth's formation, our planet underwent complex phenomena of differentiation, that slowly led to the stratigraphic division into Crust (continental and oceanic), Mantle (upper and lower), and Core (inner and outer) [27–31].

There can be no doubt that one of the basic characteristics of the early stages of the proto-Earth's formation was the presence of an atmosphere very different from today's. The origin of atmosphere and the formation of oceans are still being intensively investigated, with studies that continue to fuel scientific debate [17–19].

As can be seen from the specific literature [32–34], the most common elements are now nitrogen ($N_2 \sim 78$ %) and oxygen ($O_2 \sim 20$ %). During the planet's primordial epoch, the main constituents were H_2O and CO_2, while the accessory constituents, as is the case for the atmosphere now found on Mars and Venus, were N, CH_4 and O. What were the reasons for this relevant change? How was it possible to pass from an atmosphere that was highly toxic for present-day forms of life to one compatible with mankind's appearance on the Earth? Though these are still unanswered questions, the literature can provide us with data that suggest a number of intriguing assumptions.

First, it is interesting to note that having a primordial atmosphere saturated with CO_2 and H_2O could mean that atmospheric pressure amounted to some hundred bar, as several authors have affirmed [23–25]. In particular, the hypotheses regarding the concentrations of the various elements making up the primordial atmosphere make it possible to evaluate the effect exerted by each gas in terms of atmospheric pressure on the surface of the protocrust. These considerations are particularly important in the light of the first tectonic movements affecting Earth's life.

On the basis of the findings presented in [8, 9] and the data given in Table 13.2, reaction (13.2) would result in an overall increase in Mg of 3.2 %. This increase in Mg, however, was involved in a further piezonuclear reaction in which Mg is the starting element and C the resultant:

$$Mg_{12}^{24} \rightarrow 2C_6^{12}. \tag{13.8}$$

This hypothesis is borne out by the idea that Earth's atmosphere in the Hadean and Archean periods was very different than it is today, the concentration of C in the form of CO_2 and CH_4 being particularly high.

Different quantitative considerations can be advanced. Today, Earth's atmosphere consists of a 40 km layer, while its mass is 5×10^{18} kg with a density γ of around 1.250 kg/m^3. Given this mass and a terrestrial surface area of 5.10×10^{14} m^2, we thus have an atmospheric pressure of 96,060 Pa or 0.96 atm (~1 atm).

A similar calculation can be made for the primordial atmosphere with high concentration of Carbon. In this case, the mass to be considered will be that resulting from the piezonuclear reaction (13.8) and corresponding to 3.2 % of the mass of the protocrust involved in the reaction, $(3.4 \times 10^{21}$ kg). Considering the same gravitational acceleration for the proto-Earth, we obtain an atmospheric pressure of 666.70 atm. Although this value is very high in comparison with the current conditions of the planet's atmosphere, it can be regarded as plausible and consistent with the conditions prevailing between 3.8 and 2.5 billion years ago. In addition, it is corroborated by the studies of a number of authors who present models of a CO_2 and CH_4-rich early atmosphere with a pressures of some hundred atm, as was the case for the ground level atmospheric pressure of 660 atm indicated by Liu [17].

The large increase in light molecular composites such as O_2, CO_2, CH_4, and N_2 can be explained by the piezonuclear reactions discussed in recent papers by Carpinteri et al. [8–16]. In particular, the reactions involving Mg, Si and Al as starting elements and C, N, O, and H as resultants [8, 9]:

$$Mg_{12}^{24} \rightarrow O_8^{16} + 4H_1^1 + 4 \text{ neutrons}, \tag{13.9}$$

$$Si_{14}^{28} \rightarrow 2N_7^{14}, \tag{13.10}$$

$$Si_{14}^{28} \rightarrow C_6^{12} + O_8^{16}, \tag{13.11}$$

$$Al_{13}^{27} \rightarrow C_6^{12} + N_7^{14} + 1 \text{ neutron}, \tag{13.12}$$

can be considered to obtain a new explanation revealing the composition of the primordial atmosphere. Recent data have shown that CO_2 and H_2O concentrations increased dramatically between 3.8 and 2.5 Gyr ago (Fig. 13.1) in correspondence to sporadic seismic activity and the subduction of the primordial plates. Successively, between 2.5 and 2.0 Gyr ago, the rapid increase in N and O concentrations may be considered to be closely related to the most relevant formation of the Earth's continental crust (Fig. 13.1). The oxygen level, in fact, was very low during the Hadean and Archean eras (4.5–2.7 Gyr ago), whereas it increased sharply (Great Oxidation Event) between 2.7 and 2.4 billion years ago, until the present concentration ~21 % of the Earth's atmosphere was reached, a 10^5-fold higher value than that of the oxygen concentration in the earlier atmospheres (Fig. 13.1).

During the same period (from 2.7 to 2.4 Gyr ago), which represents 6 % of the entire Earth's lifetime (4.57 Gyr), 50 % of the continental crustal volume formed in concomitance to the most intense tectonic and continental subduction activities. Considering the scenario of approximately 2.0 Gyr ago, it is also possible to justify the origin of the first aerobic bacteria and multicellular eukaryotic organisms, ancestors of animals and human beings. Based on the composition time variations in the Earth's atmosphere shown in Fig. 13.1, and the piezonuclear fission reactions (13.6, 13.8, 13.9, 13.10, and 13.11), it can be observed that the intense tectonic activity caused a sudden increase in CO_2 and later in the O and N levels of the Earth's atmosphere. While the O and N levels remained constant, the high CO_2 concentration, principally due to piezonuclear reaction (13.8), started to decrease after 2.5 Gyr ago (Fig. 13.1). This fact can be explained considering the planetary air leak of gaseous elements and molecules, such as H, He, and CO_2, that has

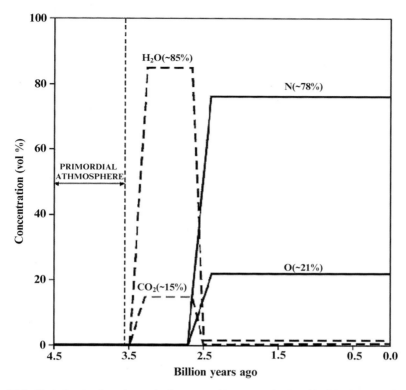

Fig. 13.1 Variation in the athmospheric composition over the Earth's life time. During the Achean era, about 15 % of the Earth's athmosphere was constituted by CO_2, and remaining part was mainly composed of H_2O. The Earth's athmospherical composition is dominated by nitrogen (N ~78 %) and oxygen (O ~21 %) [17]

affected the atmosphere during the Earth's lifetime [35]. The continuous decrease in CO_2 concentration over the last 3.5 Gyr has recently been contrasted by the well-known Carbon pollution.

The carbon dioxide (CO_2) variations over the last 500 million years (Phanerozoic period) are reported in Fig. 13.2a. The trend shown in the plot is determined considering Royer's database [36] together with the results of several specific geochemical models [37–39]. The present CO_2 concentration in the Earth's atmosphere is about 400 ppm, whereas 500 million years ago it was about 6000 ppm. Observing the phenomenon at that time scale, the increase that has occurred over the last 100 years, and which has been ascribed to Carbon emissions from the industrial revolution, seems to be negligible. In Fig. 13.2b, it is possible to observe the carbon dioxide (CO_2) variations over the last 4×10^5 years [31, 40, 41, 44, 46]. Today, some scientists sustain that, throughout the Twentieth Century, new forms of carbon pollution and the reactive nitrogen released into terrestrial environment by human activities (synthetic fertilisers, industrial use of ammonia) have

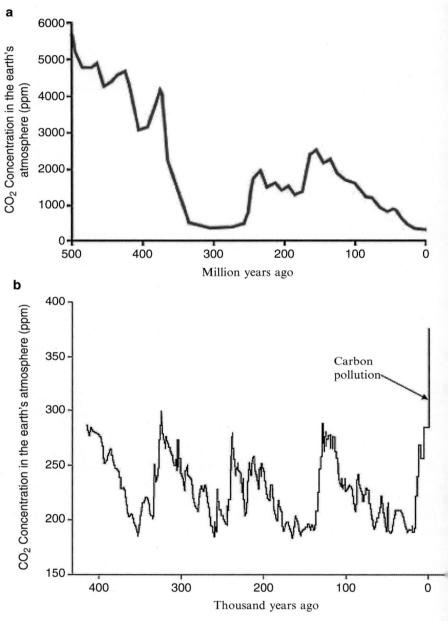

Fig. 13.2 CO_2 concentration in the Earth's atmosphere over the last 500 million years (**a**). The trend shown in the plot has been determined considering the database reported by Royer [36] together with the results of some geochemical models developed between 2001 and 2004 [38–40]. CO_2 concentration in the Earth's atmosphere over the last 4×10^5 years (**b**) [38, 39]

been the only responsible for the dramatic increase in CO_2 and N of the contemporary Earth's atmosphere [42–44]. However, a strong doubt remains whether much of these CO_2 (about 3/4 of the total) and N (about 2/3 of the total) are provided by the same causes that have produced the previous cycles of carbon dioxide concentrations (Fig. 13.2b) and the nitrogen increase in the Archean atmosphere (Fig. 13.1) [18, 32].

This hypothesis would appear to be supported by the evidence presented recently for the Earth's Crust [19, 20], together with the usually accepted idea that the early atmosphere's composition derived from Mantle degassing [31, 32, 45–47].

13.3   Compositional Evolution of the Earth Convective Mantle

Studies of the mantle's compositional evolution [32, 34, 48–50] have made it possible to analyze changes in the mass percentage concentrations of its main chemical elements and compare them with those that have been recently determined for the Earth's Crust by Carpinteri et al. [8–11].

These analyses are the basis for the graph reported in Fig. 13.3 and dealing with the changes in the mass concentrations of Fe, Mg, Si, Al, and Ca over the Earth's Mantle evolution. They lead to a number of considerations concerning the tectonic and convective events that occurred in the late Archean and their contributions to the chemical evolution of the terrestrial Mantle.

As has been observed for the Earth's Crust [8, 9], major changes in mass concentrations took place at the time of the great tectonic events, transforming heavier (Fe) into lighter (Al, Si, Mg) elements. The data presented in the literature indicate that the Earth's Mantle, like the continental crust [8, 9], saw two abrupt transitions in the concentrations of its constituent elements [32]. Specifically, these periods date between 4.0 and 3.8 billion years ago, and between 2.8 and 2.5 billion years ago. As shown in Fig. 13.3, the Mantle's decrease in Fe is balanced by the increase in other elements such as Si, Mg, Al, and Ca, which may have been involved in piezonuclear reactions like those observed for the Earth's Crust.

The data for FeO and Fe_2O_3 given in [32] indicate that Iron dropped by about 6 % at the first piezonuclear jump, and that the piezonuclear reactions (13.1 and 13.2) and a further reaction:

$$Fe_{26}^{56} \rightarrow Ca_{20}^{40} + C_6^{12} + 4 \text{ neutrons} \qquad (13.13)$$

can explain the increase of ~2.8 % in Si, ~2.4 % in Mg, ~0.3 % in Al, and ~0.35 % in Ca. These contributions are surprisingly precise in balancing the reduction in Fe in terms of mass at the Mantle. A similar situation occurred in the second critical period. Between 2.8 and 2.5 billion years ago, the decrease in Fe is again around 6 %, while the increases of ~2.3 % in Si, ~3.0 % in Mg, ~0.5 % in Al, and ~0.35 %

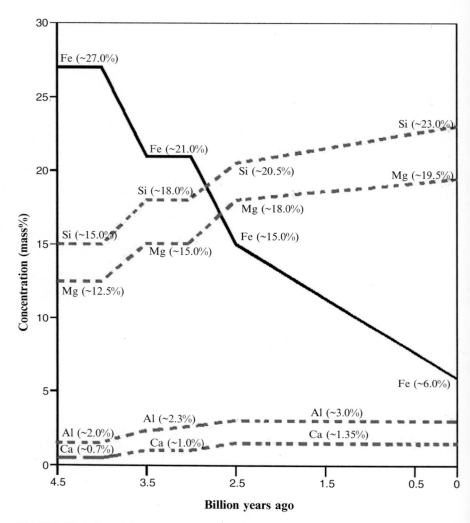

Fig. 13.3 Evolution of the concentrations of Fe, Al, Si, Mg and Ca in the terrestrial mantle over the Earth's lifetime [32]

in Ca appear to balance the reduction in Fe almost perfectly (see Fig. 13.3). The reaction involving Fe as the starting element and Ca together with C as resultants not only accounts for the increase in Ca in the mantle, but could also offer an alternative and compelling explanation for the formation of C and its anomalous concentration in the Mantle [32]. The current theories sustaining that Carbon was transported into the Mantle through the subduction of the Earth's Crust, in fact, do not seem to be entirely convincing [32].

The nearly perfectly matched balance taking place between iron and the elements increasing their concentration in both geological critical periods is surprising, and may be explained by means of the piezonuclear reactions described above.

However, it is necessary to assume that the Earth's Mantle, unlike its Crust, went through a further change in concentration, that involved the same elements analyzed earlier, from 2.5 billion years ago up to the present time (Fig. 13.3).

In this last change, we can see that the elements' concentrations have varied continuously and not step-wisely, and that these variations were thus not confined to a very narrow span of time as they were in the previous cases. From 2.5 billion years ago up to the present time, there has been a further decrease in the concentration of Fe amounting to approximately 11.5 %, and increases in other elements (in particular, Si and Mg) which have only partially balanced this reduction [32]. Indeed, significant increases can be observed in Si (+2.4 %) and Mg (+1.6 %), with an overall increase of only 4 %.

Whereas the remaining not balanced decrease in Fe is approximately equal to 7.5 % of the Mantle's mass. This decrease could be explained considering the formation of the Earth's Core. This percentage of iron may have gone to making up part of the Outer Core, where the concentration of iron is approximately of 80 % [8, 9, 28, 32].

13.4 Conclusions

The main results of the present paper are related to the evidence of compositional changes which affected the Earth's Crust, Atmosphere, and Mantle, from the early period of their evolution until today. It was observed how some anomalies in the concentrations of the most abundant elements in the Mantle can be interpreted by the piezonuclear fission reactions already introduced by Carpinteri and his team. In addition, it was observed that a further reaction involving Fe, Ca, and C may be assumed to comprehend the Earth's Mantle evolution. This reaction not only accounts for the increase in Ca at the Mantle, but could also offer an alternative and compelling explanation to the formation of Carbon and to its anomalous concentration in the Mantle [32]. The current theories sustaining that Carbon was transported into the Mantle through the Crust's subduction, in fact, do not seem to be entirely convincing [32]. From 2.5 billion years ago up to the present time, there has been a further decrement in the concentration of iron amounting approximately to 11.5 %, as well as related increments in other elements (in particular, Si and Mg) which have only partially balanced this reduction [32]. Indeed, significant increments can be observed in Si (+2.4 %) and Mg (+1.6 %), with an overall increase of only 4 %. The decrease in Fe which is not balanced by the increases in these two elements (Si and Mg) amounts to 7.5 % of the total Mantle's mass. This decrement could be related to the formation of the Earth's Core, as this percentage of iron may have gone to making up part of the Outer Core, where the concentration of iron is approximately of 80 % [8, 9, 28, 32].

In this context many hypotheses have been proposed to explain the time lag. Some of the most known are the Tectonic Trigger [50], the Nickel famine [5] and the Bistability model [51]. According to these hypothesis is very difficult to determine a quantitative evaluation about the causes producing the Oxygen concentration 2.5 billion years ago and today. Among these explanations the proposed conjecture based on tectonic activity and piazonuclear reaction seems to be useful under the light of a scientific debate devoted to clarify the causes producing the Earth oxygenation.

References

1. Mao HR, Hemley J (2007) The high-pressure dimension in earth and planetary science. Proc Natl Acad Sci USA 104:914–915
2. Anbar AD (2008) Elements and evolution. Science 322:1481–1482
3. Taylor SR, McLennan SM (2009) Planetary crusts: their composition, origin and evolution. Cambridge University Press, Cambridge
4. Saito MA (2009) Less nickel for more oxygen. Nature 458:714–715
5. Konhauser KO et al (2009) Oceanic nickel depletion and a methanogen famine before the great oxidation event. Nature 458:750–754
6. Favero G, Jobstraibizer P (1996) The distribution of aluminum in the Earth: from cosmogenesis to Sial evolution. Coord Chem Rev 149:367–400
7. Taylor SR, McLennan SM (1995) The geochemical evolution of the continental crust. Rev Geophys 33:241–265
8. Carpinteri A, Manuello A (2011) Geomechanical and geochemical evidence of piezonuclear fission reactions in the Earth's crust. Strain 47:282–292
9. Carpinteri A, Manuello A (2012) An indirect evidence of piezonuclear fission reactions: geomechanical and geochemical evolution in the Earth's crust. Phys Mesomech 15:14–23
10. Carpinteri A, Borla O, Lacidogna G, Manuello A (2010) Neutron emissions in brittle rocks during compression tests: ,monotic vs. cyclic loading. Phys Mesomech 13:264–274
11. Carpinteri A, Lacidogna G, Manuello A, Borla O (2011) Energy emissions from brittle fracture: neutron measurements and geological evidences of piezonuclear reactions. Strength Fract Complex 7:13–31
12. Carpinteri A, Lacidogna G, Manuello A, Borla O (2012) Piezonuclear fission reactions: evidences from microchemical analysis, neutron emission, and geological transformation. Rock Mech Rock Eng 45:445–459
13. Carpinteri A, Lacidogna G, Manuello A, Borla O (2013) Piezonuclear fission reactions from earthquakes and brittle rocks failure: evidence of neutron emission and nonradioactive product elements. Exp Mech 53(3):345–365
14. Cardone F, Carpinteri A, Lacidogna G (2009) Piezonuclear neutrons from fracturing of inert solids. Phys Lett A 373:4158–4163
15. Carpinteri A, Cardone F, Lacidogna G (2009) Energy emissions from failure phenomena: mechanical, electromagnetic, nuclear. Exp Mech 50:1235–1243
16. Carpinteri A, Cardone F, Lacidogna G (2009) Piezonuclear neutrons from brittle fracture: early results of mechanical compression tests. Strain 45:332–339, Atti dell'Accademia delle Scienze di Torino, Torino, Italy, 33:27–42
17. Liu L (2004) The inception of the oceans and CO_2-atmosphere in the early history of the Earth. Earth Planet Sci Lett 227:179–184
18. CRC Handbook of Chemistry and Physics (1980) Robert C. Weast (ed). CRC Press, New York, F-199

19. Garrison TS (2005) Oceanography: an invitation to marine science. Thompson Brooks Cole, Belmont
20. Schopf J (1983) Earth's earliest biosphere: its origin and evolution. Princeton University Press, Princeton
21. Kolb E (2000) Blind watchers of the sky: the people and ideas that shaped our view of the universe. Oxford University Press, Oxford
22. Kolb E, Matarrese S, Notari S, Riotto A (2005) Primordial inflation explains why the Universe is accelerating today. arXiv:hep-th/0503117v1, 1–4
23. Williams RP, Da Silva FJR (2003) Evolution was chemically constrained. J Theor Biol 220:323–343
24. Buesseler KO, Doney SC, Karl DM et al (2008) Ocean iron fertilization moving forward in a sea of uncertainty. Science 319:162
25. Holland HD (2006) The oxygenation of the atmosphere and oceans. Philos Trans R Soc Lond Ser B 361:903–915
26. Abbott DH, Burgess L, Longhi J, Smith WHF (1994) An empirical thermal history of the Earth's upper mantle. J Geophys Res 99(13):835–850
27. Vovna GM, Mishkin MA, Sakhno VG, Zarubina NV (2009) Early archean sialic crust of the Siberian craton: its composition and origin of magmatic protoliths. Dokl Earth Sci 429 (2):1439–1442
28. The World Ocean (2007) The Columbia encyclopedia. CD-ROM, 6th edn. Columbia University Press, New York
29. Van Nostrands Scientific Encyclopedia (2008) Ocean volume and depth, 10th edn. Van Nostrands Scientific Encyclopedia, New York
30. (1994) The concise Columbia electronic encyclopedia, 3rd edn. Columbia University Press, New York
31. Kasting JF, Ackerman TP (1986) Climatic consequences of very high carbon dioxide levels in the Earth's early atmosphere. Science 234:1383–1385
32. Yung YL, De More WB (1999) Photochemistry of planetary atmospheres. Oxford University Press, New York
33. Ronov AB, Yaroshevsky AA (1978) The chemical composition of Earth's crust and its shells. In: Tectonosphere of Earth. Nedra, Moscow, pp 376–402
34. Catling CD, Zahnle KJ (2009) The planetary air leak. Sci Am 300:24–31
35. Royer DL, Berner RA, Montanez IP, Tabor NJ, Beerling DJ (2004) CO2 as a primary driver of Phanerozoic climate. GSA Today 14:4–10
36. Berner RA (1990) Atmospheric carbon dioxide levels over Phanerozoic time. Science 249:1382–1386
37. Berner RA, Kothavala Z (2001) Geocarb III: a revised model of atmospheric CO_2 over Phanerozoic time. Am J Sci 301:182–204
38. Bergman RA, Noam M, Timothy ML, Watson AJ (2004) Copse: a new model of biogeochemical cycling over Phanerozoic time. Am J Sci 301:182–204
39. Rothman DH (2001) Atmospheric carbon dioxide levels for the last 500 million years. Proc Natl Acad Sci USA 99:4167–4171
40. Fischer H, Wahlen M, Smith J, Mastroianni D, Deck B (1999) Ice core records of atmospheric CO_2 around the last three glacial terminations. Science 283:1712–1714
41. Monnin E, Steig EJ, Siegenthaler U et al (2004) Evidence for substantial accumulation rate variability in Antarctica during the Holocene, through synchronization of CO2 in the Taylor Dome, Dome C and DML ice cores. Earth Planet Sci Lett 224:45–54
42. Townsend A, Howarth RW (2010) Fixing the global nitrogen problem. Sci Am 302:50–57
43. Aki K (1983) Strong motion seismology. In: Kanamori H, Boschi E (eds) Earthquakes: observation, theory and interpretation. North-Holland Pub Co, Amsterdam, pp 223–250
44. Sorokhtin OG et al (2007) Global warming and global cooling – evolution of climate on Earth. Elsevier, Amsterdam
45. Ahrens TJ (1971) The state of mantle minerals. Technophysics. XX:189–219

46. Wang L et al (2010) Nanoprobe measurements of materials at megabar pressures. Proc Natl Acad Sci USA 107(14):6140–6145
47. Rngwood AE (1962) The chemical composition and the origin of Earth. In: Hurley PM (ed) Advance in earth science. MIT PRESS, Cambridge
48. Urey HC, Craig H (1953) The composition of the stone meteorites and the origin of the meteorites. Geochim Cosmochim Acta 4(1–2):36–82
49. Rees M (2005) Universe – the definitive visual guide. Dorling Kindersley Ltd, NewYork
50. Lenton TM, Schellnhuber HJ, Szathmáry E (2004) Climbing the co-evolution ladder. Nature 431(7011):913
51. Goldblatt C, Lenton TM, Watson AJ (2006) The great oxidation at 2.4 Ga as a bistability in atmospheric oxygen due to UV shielding by ozone. Geophys Res Abstr 8:00770

Chapter 14
Piezonuclear Fission Reactions Triggered by Fracture and Turbulence in the Rocky and Gaseous Planets of the Solar System

Alberto Carpinteri, Amedeo Manuello, and Luca Negri

Abstract Evidences from the planets of the Solar System are presented and interpreted in the light of piezonuclear fission reactions. In particular, results coming from different investigations are reported for the crust of Mars. They were made available by NASA space missions during the last 15 years. The concentration increment in certain elements (Fe, Cl, and Ar) and the corresponding decrement in others (Ni and K), together with neutron emissions at Mars largest faults, should be considered as directly correlated phenomena. The findings presented provide a clear evidence of how seismic activity has contributed to the Red Planet's compositional evolution. Analogous evidences regard Mercury, Jupiter, and the Sun itself. The major compositional variations are interpreted according to piezonuclear fission reactions triggered by earthquakes in rocky planets and by storms in gaseous planets as well as in our star. These conjectures, which were originated from the analysis of geological and geophysical evolution of the Earth's crust, are based on recent evidence of neutron and alpha particle emissions during brittle fracture experiments carried out on inert non-radioactive rocks (granite, basalt, magnetite, marble).

Keywords Chemical evolution • Fracture • Turbulence • Rocky planets • Gaseous planets • Solar system

14.1 Introduction

The paper will present evidences from the literature on the cosmology of the Solar System [1–11], with particular emphasis on the chemical composition evolution of Mars [1–3, 6–8]. The discussion will focus onto the Red Planet's crust studies made possible by NASA space missions during the last 15 years, including the Mars Odyssey (2001) and Mars Pathfinder (1996) missions [5]. Data about the compositional analysis of the planet's crust are correlated to maps of the major faults

A. Carpinteri (✉) • A. Manuello • L. Negri
Department of Structural, Geotechnical and Building Engineering, Politecnico di Torino, Corso Duca degli Abruzzi 24, 10129 Torino, Italy
e-mail: alberto.carpinteri@polito.it

© Springer International Publishing Switzerland 2015
A. Carpinteri et al. (eds.), *Acoustic, Electromagnetic, Neutron Emissions from Fracture and Earthquakes*, DOI 10.1007/978-3-319-16955-2_14

produced by altimetric surveys, and of the neutron emission data provided by the HEND High Energy Neutron Detector carried on Mars Odyssey 2001, which used He^3 proportional counters to measure epithermal neutrons (energy range: 0.4 eV– 100 keV) [3, 4, 11]. The concentration increment in certain elements (Fe, Cl, and Ar), the corresponding decrement in others (Ni and K), and the emission of neutrons at Mars largest faults should be considered as correlated phenomena [1–3, 6–8, 11]. The chemical composition evolution of other planets in the Solar System will also be investigated on the basis of the data accumulated in a large number of space missions, Mariner 10, Messenger (2004) for the Sun and for Mercury and Venus, Mars Odyssey (2001), Pioneer 11 (1973), Voyager 1 (1977), and the Galileo probe (1989) for the outer planets Jupiter and Saturn. The hypothesis of piezonuclear fission reactions will be presented as a new key for the interpretation of certain astrophysical phenomena which are associated to the chemical composition of each heavenly body and still lacking in a clear explanation [12–22]. The planets can be classified into two categories: the terrestrial, earth-like, or rocky planets (Mercury, Venus, Earth, and Mars), and the gaseous planets (Jupiter, Saturn, Uranus, and Neptune) [1]. There are several differences between the two groups of planets. First, the rocky planets all have a small mass, few or no moons, and a low rotational speed, whereas the gaseous planets have a large mass, several moons, and a high rotational speed. For this reasons, the gaseous planets are flatter at the poles than the rocky planets. In addition, the latter's density is five times that of water on average, whereas the density of the gaseous planets is only 1.2 times that of water [1]. In examining their composition, it has been seen that the terrestrial planets consist essentially of rocky and metallic materials; by contrast, the gaseous planets are mostly made up of helium, hydrogen and small quantities of ice [1]. Furthermore, the terrestrial planets either have no atmosphere at all or one that is in any case rarefied, by contrast with the gaseous planets' dense atmosphere consisting of hydrogen, helium, ammonia, phosphine, and methane [1]. In particular, the earth-like planets differ from the gaseous planets in their internal structure. The former have a solid outer crust, a mantle, and a liquid inner core, whereas the latter are called "gas giants" precisely because they consist almost entirely of volatile elements. As has already been done for the Earth, we apply the hypothesis of piezonucear fission reactions to reconstruct the compositional evolution of the other planets, as well as of the Sun itself. The planets that have been investigated so far belong to the rocky category (Earth and Mars), with the phenomenon of piezonuclear fission being triggered by seismic activity, as well as to the gaseous ones (Jupiter and Saturn), where the piezonuclear reactions are triggered by turbulence. The findings presented herein provide further evidence of how fracture and flow instability can have contributed to the planets' compositional evolution, as they did for Earth [10, 17, 22]. The major compositional variations were generated by high frequency pressure waves in both cases [12–22]. These conjectures, which have been originated from an analysis of the geological and geophysical evolution of the Earth's crust, are based on recent experimental evidence of neutron emissions during brittle fracture tests carried out on specimens of inert non-radioactive rocks (granite, basalt, magnetite) [12–15, 18–21].

In the last few years, neutron emissions have been observed in several experiments characterized by static or fatigue tests on inert rocks [12–22]. These evidences suggest that high-frequency pressure waves generate anomalous nuclear reactions, producing energy emission in the form of neutrons [12–14].

At the same time, theoretical interpretations were proposed. Cardone and coworkers [23, 24] have pursued a theoretical explanation based on deformed spacetime. Widom et al. explained neutron emissions as a consequence of nuclear fission reactions taking place in iron-rich rocks during brittle micro-cracking or fracture [25, 26]. Several evidences are observed that iron nuclear disintegration takes place when rocks containing such nuclei are crushed and fractured. The resulting nuclear trasmutations are particularly relevant in the case of magnetite and iron-rich materials in general. The same authors argued that neutron emissions may be related to piezoelectric effects and that the fission of iron may be a consequence of photodisintegration of the nuclei [25, 26]. At the same time Hagelstein and Chaudhary reported that the anomalous reactions may be considered under the light of specific conditions in fracture experiment with large amplitude excitation of vibrational modes considering phonon-electron coupling off of resonance [27]. The anisotropic and asynchronous neutron pulses observed at the time of fracture were accompanied by direct signs of compositional changes, as they were later observed on the specimens' fracture surfaces under the electron microprobe [16]. These changes observed in the laboratory were also confirmed from analyses of the composition evolution in the Earth's crust [17, 22]. This interpretation of our planet's geophysical and geological evolution is corroborated by the locations of the mineral reservoirs on the Earth's surface, as well as by the variations in the composition of the continental crust that have taken place over the last 4.5 Billion years [17, 22].

The studies of the planet Mars presented here to some extent follow the same approach used in analyzing the compositional evolution of the Earth's crust and demonstrate how, for the Red Planet as for our own, it is possible to reinterpret the changes over time in the main elements making up its surface crust (Fe, Ni, K, and Cl in particular), as well as the effects on the concentration of argon, which in the form of its isotope Ar^{36} is particularly common in the Martian atmosphere [1, 3, 4, 8, 11].

14.2   Internal Structure and Chemical Composition of Mars

Mars was formed when rocky fragments that remained in the nebula immediately after the formation of the outer gas giants clustered together [1]. As a result of their formation, the so-called rocky planets are nearly devoid of any volatile elements [1]. In comparison with the other planets, Mars has almost twice the quantity of volatile elements, a higher level of oxidation, a smaller core and a primordial core composition with around twice as much iron as that of the Earth. With all of these characteristics, Mars has differed substantially from the Earth since the time of its formation 4.57 Billion years ago [1].

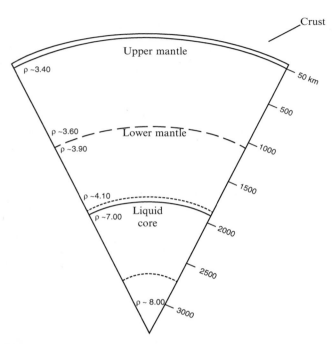

Fig. 14.1 Model of Mars' internal structure, showing crust, mantle and core with the associated uncertainties for mantle and core thickness. Densities are expressed in g/cm³ [1]

Today, Mars' stratigraphy and the actual thicknesses of its crust, mantle, and core are still relatively unclear, as there are no data from seismic analyses and they have thus been described on the basis of mathematical models that are affected by uncertainties [1]. Thanks to the findings of the 1997 Pathfinder mission, however, the data now available on the planet's stratigraphy and density provide a picture that is sufficiently detailed to permit realistic hypotheses [5].

Mars has an average radius of 3389.9 km and a volume of 1.6317×10^{11} km³. Its mass is 6.4185×10^{23} kg, giving it a total density of 3.934 g/cm³ (see Fig. 14.1).

The thickness of Mars' crust is highly variable, and ranges from a minimum of 20 km to a maximum of 70 km, with a predominantly basaltic composition [1]. An accurate evaluation of the current average composition of Mar's crust surface is shown in Table 14.1. The differences between the composition of the Red Planet's crust and that of the Earth will be discussed below. As indicated in Fig. 14.1, the average density of Mars' crust alone is around 3.0 g/cm³, approximately 16 % more than the density of Earth's present continental crust (2.5 g/cm³). The density of Mars' crust is close to the estimated density of the terrestrial protocrust during the Archean period between 3.8 and 2.5 Billion years ago (3.2 g/cm³) [17, 22]. Considering the Solar System's three types of primary, secondary, and tertiary crust described by Taylor and McLennan [1, 17, 22], Mars' crust can be classified as a secondary crust resembling our planet's oceanic crust or the Lunar Seas, and thus is

Table 14.1 Average quantities of the major elements constituting Mars' crust

Element	Concentration (mass %)	Element	Concentration (ppm)
Si	23.0	Ti	5880
Fe	15.0	P	3930
Al	5.5	K	3740
Mg	5.4	Mn	2790
Ca	4.9	Cr	2600
Na	2.2	Ni	337

Concentrations are shown in mass percentage or ppm

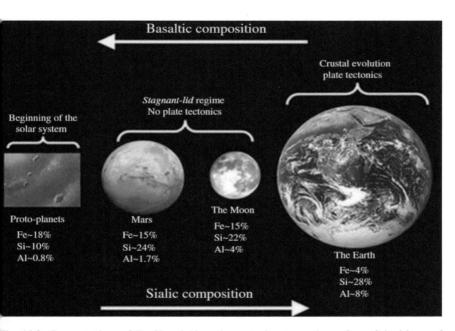

Fig. 14.2 Concentrations of Fe, Si and Al on the protoplanets, on the surface of the Moon, of Mars, and on Earth's crust. Note that the concentration of Fe drops while the concentrations of Si and Al increase in passing from the first formations making up the protoplanets to the current status of the planet Earth

at a stage of evolution midway between primary and tertiary crust [1, 17, 22]. In view of the hypothesis regarding planetary-scale evolution presented by Carpinteri and Manuello [17, 22], we can understand how Mars, which like the Earth saw an extremely intense tectonic activity during the early stages of its formation, is on an intermediate step of the evolutionary scale based on the surface composition of the Solar System's planets, or, in other words, between the original bodies (proto-planets) and the present-day, chiefly sialic composition of the tertiary crusts, whose only known example is the Earth continental crust (see Fig. 14.2).

Figure 14.3 shows the concentrations expressed in mass percentages of the elements making up the core, the mantle, and the crust surface of Earth and Mars, while the same data are given in Tables 14.1 and 14.2 for the Red Planet. All data indicated are taken from the specialist literature [1, 28–33] and point to several basic differences between the two planets. First, the current quantity of Fe present on the surface of Mars, which accounts for 15 % of the planet's crust, is around 1.8 times the amount of Fe present in the Earth's oceanic crust (~7.8 % by mass) and around 3.5 times that in the continental crust (~4.0 %) [1, 9, 10, 22] (Fig. 14.3c and Table 14.1). About silicon, the concentration on

Mars (23 %) is approximately 17 % lower than that found in Earth's continental crust (28 %), while the situation for Al is similar, as the 5.5 % concentration on Mars is some 30 % less than the approximately 8 % concentration in the continental crust. The values of Mg, Ca, and Na are also significant. Mars' mantle is approximately 1050 km thick, and is divided into an upper mantle and a lower mantle with different specific weights. That of the upper mantle ranges from a minimum of 3.4 g/cm^3 at a depth of around 50 km up to a maximum of 3.6 g/cm^3 at 1000 km, while the lower mantle starts from a density of 3.9 g/cm^3 at a depth of 1050 km and arrives at the core-mantle boundary layer with an average density of 4.1 g/cm^3. As mentioned earlier, these densities are determined from studies carried out with mathematical models rather than measured in direct analyses; hence, for instance, the uncertainty regarding the position of the boundary between the lower mantle and the core, which varies over a range of around 500 km. The average chemical composition of the mantle presented in the literature, on the other hand, was determined from spectrographic analyses carried out on meteorites from Mars which landed on Earth in the primordial era. Consequently, the chemical composition indicated for the mantle refers to a period differing from that available for the crust (Table 14.2) [1].

The size, composition and state (i.e., whether it is liquid or solid) of Mars' core are still hotly debated. However, hypotheses have been advanced about what dimensions, makeup and state of aggregation are most likely in the light of the information in the literature. Mars' core has a radius of about 1600 km (allowing for the model mentioned above) and a density of 7–8 g/cm^3.

14.3 Geological History of Mars

The genesis of the Martian surface is still a subject of controversial studies, given that by contrast with our own planet, it is not possible to establish a timescale that can be considered accurate, as the only way of dating the planet is by counting the craters left behind by meteorite impacts [23, 28]. The geological history of Mars is roughly divided into three main periods: Noachian, Hesperian, and Amazonian [4]. As has been repeatedly confirmed by a number of scientists, Mars is probably no longer an intensely active planet from the tectonic standpoint, but there can be no

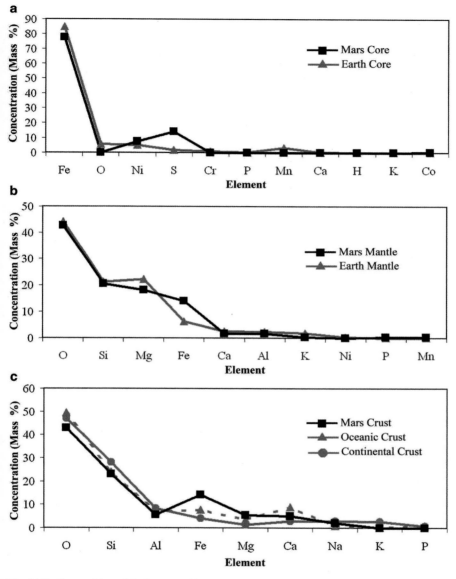

Fig. 14.3 Composition of the layers making up Earth and Mars. Average composition of the core for Earth and Mars (**a**). Average composition of the mantle for the two planets (**b**). Average composition of Earth's continental and oceanic crust and average composition of Mars' surface (**c**)

Table 14.2 Average composition of Mars' primordial mantle

Element	Concentration ppm	Element	Concentration (mass %)
Si	20.6	P	700
Fe	13.9	K	305
Al	1.6	Ti	840
Mg	18.2	Cr	5200
Ca	1.75	Ni	400
Na	0.37	Zn	62
Mn	0.36	Co	68

Concentrations are shown in mass percentage or ppm

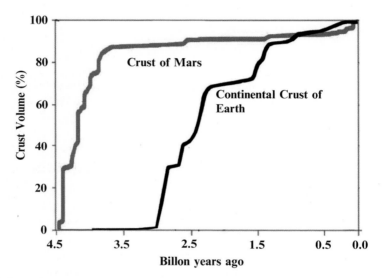

Fig. 14.4 Percentage formation of the crust of Mars and of Earth's continental crust

doubt that it was such in the past and that it continues to show a sporadic tectonic activity today [1, 4, 5, 11, 29, 30].

The fact that it does not show abundant signs of ongoing tectonic activity does not necessarily mean that Mars has always been seismically stable or lacking in internal dynamics. A look at the different epochs in which Martian geological history is divided, in fact, will show that the Noachian Period was one of dramatic tectonic and volcanic activity [1]. Over 80 % of the planet's crust was formed during this process (see Fig. 14.4) [4]. Figure 14.4 shows a comparison of the percentage formation curves for Earth's continental crust and for Mars. As can be seen, the formation of the terrestrial crust was spread over the last 3.8 Billion years, whereas the formation of Mars' crust can be reasonably dated between 4.57 and 3.50 Billion years ago. This evidence indicates that the Red Planet's tectonic activity came substantially to an end after the Noachian Period, "crystallizing" the crust's evolution with no further macroscopic changes.

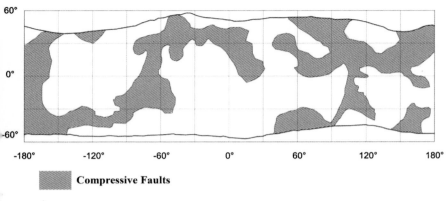

60°

0°

-60°

-180°   -120°   -60°   0°   60°   120°   180°

Compressive Faults

Fig. 14.5 Map showing the distribution of Mars faults on a planisphere of the planet

Proof that Mars' crust was the scene of major tectonic events in the past is also provided by the presence of a large number of faults on the planet's surface (Fig. 14.5) [4, 11]. Created during the Noachian Period, these faults are the result of both the planet's internal dynamics [1, 4] and of the global cooling that was followed by an overall contraction of the lithosphere [11]. Figure 14.5 is a geological map of Mars showing the location of the areas with the highest concentration of thrust faults on a planisphere of the planet [4].

14.4 Compositional Evolution of the Martian Crust and Considerations Regarding the Role of Piezonuclear Reactions

In the wake of recent studies of the chemical composition evolution of Earth's continental crust [17, 22], the major compositional changes have been interpreted in the light of piezonuclear reactions in natural materials that have been observed and confirmed in laboratory experiments [22]. The surprising findings that have emerged from analysis under the electron microprobe have highlighted how piezonuclear reactions can be used to explain the geological evolution, the composition of the proto-atmosphere, the decrease in Ni and Fe in the oceans, and many other phenomena during the life of our planet [17, 22].

Similar hypotheses can be formulated for Mars, as the violent tectonic and volcanic activity it experienced in the Noachian Period, and the sporadic tectonic events that still today take place, have resulted in a surface marked by faults and signs of earthquakes [28–30]. In this connection, mention should be made to the fact that Mars is commonly known as the "red planet" because of the characteristic red colour of its surface. This colour results from the iron oxide, FeO, that covers much of the planet's crust and is the second most common compound in it [1]. This

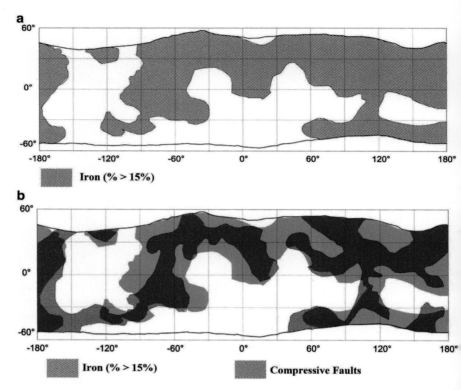

Fig. 14.6 Fe-rich areas of the planet's surface (**a**). The map given above was superimposed on the map of the faults located on the planet's surface (**b**)

large quantity of Fe – around 15 % by mass, as compared with ~7.8 % in the Earth's oceanic crust [1], which is richer in iron than the continental crust (~4.0 %) – raises interesting questions about the planet's evolution and the role that unexpected nuclear reactions may have had in this process.

Evidence of the possible role of these reactions can be provided by the data regarding the distribution of Fe on Mars' surface [1, 4, 8], which we have superimposed on the map of compressional faults in the crust published by Knapmeyer et al. [4] (Fig. 14.6a). As can be seen from Fig. 14.6b, there is a very high correlation between the areas showing extensive signs of tectonic activity and the concentration of Fe on the surface.

In recent years, numerous studies have addressed the composition of Mars' surface. In particular, some researchers have been able to reconstruct maps of the surface concentrations of elements such as H, Si, Fe, K, Cl, Th, and Si by analyzing the results provided by the Gamma Ray Spectrometer (GRS) on board of the 2001 Mars Odyssey spacecraft [8].

The exploratory Mars Odyssey mission also analyzed the neutron emission data provided by the HEND He^3 High Energy Neutron Detector carried on the

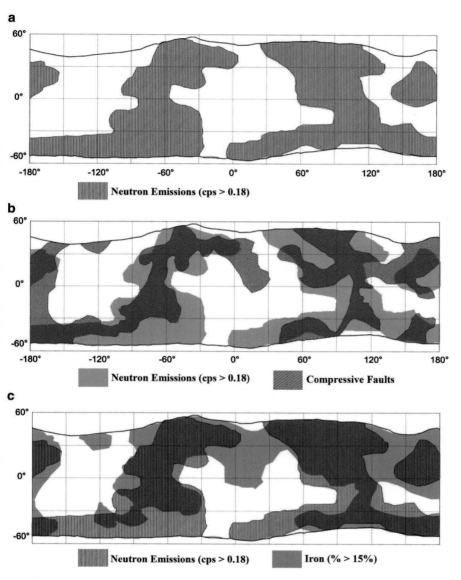

Fig. 14.7 Neutron emissions (**a**). Seismic areas and neutron emissions (**b**). The emissions map was superimposed on that for Fe concentration (**c**)

spacecraft to measure epithermal neutrons in an energy range of 0.4 eV–100 keV [2–5, 8].

Figure 14.7a shows a map of epithermal neutron emissions from the planet's surface. In Fig. 14.7b, c, such emissions are superimposed, first on the data for the location of the major faults, and then on the map of Fe concentration on the Red

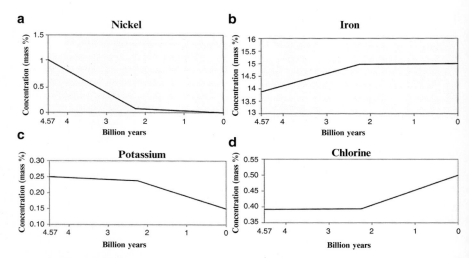

Fig. 14.8 Evolution of the concentrations of Fe (**a**), Ni (**b**), K (**c**) and Cl (**d**) in the compositions of Mars' surface

Planet's surface. In both cases, the correlations between neutron emissions, Fe concentrations and fault locations on the planetary crust are surprisingly close. Here, the underlying assumption which explains the high correlation between these data is that Fe is not the starting element due to a piezonuclear reaction – as it is on Earth, but is the resultant of reactions that involve elements with a higher atomic number as the starting elements (Ni, Sn, Cu). In the scientific literature, moreover, several studies report an increase in the concentration of Fe on the Red Planet's surface over the last 4.57 Billion years [1, 2, 8, 31–33]. At the same time, a number of authors report a sizeable decrease in elements such as Cu, Ni, Co, Mo, In, Zn, and Sn [1, 7, 31, 32].

In particular, the absolute increase in Fe can be evaluated from the approximately 14 % concentration in the Noachian crust to today's ~15 %, a relative increase of around 1 % corresponding to about 2.15×10^{20} kg of Mars' crust (Fig. 14.8a). Likewise, the data for Ni concentrations, as shown in Fig. 14.8b, indicate a roughly equivalent decrease in this element, passing from a concentration of around 1 % in the Noachian Period to a negligible or near-zero concentration in Mars' present-day crust. This is thus an absolute decrease of about 100 % and a relative decrease of 1 % [1].

As has been assumed for other elements in the case of Earth's crust, this evidence suggests that the following piezonuclear fission reaction took place:

$$Ni_{28}^{59} \rightarrow Fe_{26}^{56} + 2H_1^1 + 1 \text{ neutron} \tag{14.1}$$

This reaction could explain the anomalous increase in the Fe concentration and the corresponding decrease in Ni by the same amount in terms of mass (Fig. 14.8a, b).

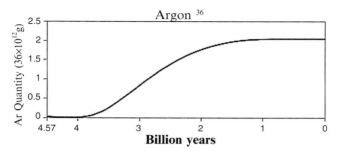

Fig. 14.9 Evolution of the quantity of Ar^{36} in Mars' atmosphere [31]

In addition, the data presented by Hahn and McLennan and by Boynton et al. [2, 8] also emphasize anomalous variations in K and Cl. A small portion of the decrease in K, moreover, can be directly correlated through piezonuclear reactions to the high concentration of an isotope of argon, Ar^{36}, in Mars' atmosphere. This evidence also makes it possible to hypothesize two further fission reactions that may have affected the planet's crust [1, 2, 8] (Fig. 14.8c, d):

$$K_{19}^{39} \rightarrow Cl_{17}^{35} + 2H_1^1 + 2 \text{ neutrons} \tag{14.2}$$

$$K_{19}^{39} \rightarrow Ar_{18}^{36} + H_1^1 + 2 \text{ neutrons} \tag{14.3}$$

By combining the data from various studies, we can see an overall decrease in K (~0.1 %) at the planet's faults and a virtually identical relative increase in Cl (~0.1 %). This evidence suggests that piezonuclear reaction (14.2) was responsible for the decrease in potassium in the Red Planet's crust reported by a number of authors. In addition, several authors have mentioned a further point of particular interest. Mars' atmosphere would appear to be laden with an anomalously high concentration of Ar^{36}, around 2.5 times higher than that in the atmosphere of Earth. The clear increase in the concentration of Ar^{36} during the planet's geological evolution seems to be correlated to tectonic activity, and may be linked to the decrease in K through reaction (14.3) (Fig. 14.9).

This reaction and the resulting decrease in K followed by an increase in Ar^{36}, though not particularly important for the balances and compositional changes that can be observed in the planet's crust, would appear to be especially interesting as regards the evolution of Mars' proto-atmosphere and its present composition.

14.5 The Others Planets and the Sun

Similar evidence can be seen for other bodies in the Solar System. The Earth-like heavenly bodies differ from those referred to as gas giants in their internal structure. The former consist of a solid crust, a mantle, and a liquid outer core, whereas the gaseous plants are composed almost entirely of volatile elements.

As was done for Earth and Mars, we have attempted to reconstruct the compositional evolution of other planets and the Sun on the basis of the data found in the literature, using the piezonuclear hypothesis to shed light onto these still inadequately understood changes.

Up to this point, piezonuclear fission has been discussed in connection to the compositional evolution of the planet's layers and anomalous neutron emissions for both Earth and Mars. For those planetary bodies that have no solid crust (Jupiter and Saturn) and thus cannot be affected by tectonic phenomena, a different approach in keeping with their state of aggregation will now be used.

14.5.1   Considerations Regarding the Chemical Composition of Mercury's Crust and Atmosphere

Mercury is the planet closest to the Sun, orbiting at a distance of 57.9 Million km as opposed to the Earth's 149.6 Million km [1, 34]. Because of the high temperatures of the planet's surface and the major influence that the solar wind has on it, it was not possible in the past to perform in-depth, accurate analyses of the composition of Mercury's crust and atmosphere. The only certain data that have been obtained were chiefly provided by two space missions: Mariner 10 (1974–1975) and Messenger (2004–2011). The latter probe launched by NASA to analyze the planet's exosphere, magnetosphere, surface, and interior was equipped with an X-Ray spectrometer capable of measuring the quantities of elements such as Mg, Al, Ca, Si, Fe, and S on Mercury's surface [35]. In the past, Mercury was always thought of as the Moon's twin, as its basaltic surfaces were believed to be similar in structure and composition. The first results of an analysis of the composition of Mercury's surface and the chemical composition of lunar basalts showed that Mercury's crust is richer than the crust of the Moon in elements such as Mg and S but has lower concentrations of other elements such as Al, Fe, and Ca [1].

Unlike Earth, Mercury has no plate tectonics. This does not mean that the planet does not have, or did not have in the past, some form of tectonic activity. In fact, the observations made with the MDIS (Mercury Dual Imaging System) module on the Messenger probe, which takes high-definition photographs of the planet's surface, have provided abundant evidence of Mercury's past tectonic activity. In particular, it has been noted that the planet's surface is marked by numerous scalloped cliffs called lobate scarps caused by contraction of the planet's crust accompanying cooling of its core [36]. This contraction, which doubtless took place in the past and is almost certainly still under way today [36], results in enormous compressive stresses in the plane tangent to the surface, forcing the latter upwards to form the lobate scarps [36]. Mercury has a very weak atmosphere consisting of traces of H, He, O [1] and unexpected concentrations of Na ions [37]. There is still no consensus regarding the phenomenon behind the presence of sodium ions in Mercury's exosphere, as there are several contrasting theories for its cause [37]. An alternative explanation, associated with the planet's tectonic activity, is provided by piezonuclear reactions. In observing the data obtained by Messenger on the composition of Mercury's crust [35], which is poor in elements such as Si, Al, and Ca, and of the Na ion-rich atmosphere [37], it can be assumed that the following piezonuclear reactions took place:

$$Al_{13}^{27} \rightarrow Mg_{12}^{24} + H_1^1 + 2 \text{ neutrons} \tag{14.4}$$

$$Al_{13}^{27} \rightarrow Na_{11}^{23} + He_2^4 \tag{14.5}$$

$$Ca_{20}^{40} \rightarrow S_{16}^{32} + 2He_2^4 \tag{14.6}$$

Through these reactions, it can be seen that the decrease in elements such as Si and Al is accompanied by an increase in Mg, which is common in the crust, and Na, which Mercury's atmosphere contains in anomalous quantities.

14.5.2 Kelvin-Helmholtz Instability and Its Role in the Compositional Evolution of Jupiter and Saturn

Kelvin – Helmholtz instability is a type of turbulent flow instability that arises when different layers of fluid are in motion relative to each other. It was discovered and investigated by Lord Kelvin [38] and by Helmholtz [38], and later by Rayleigh. The simplest example that can be imagined in two dimensions is that of a perfect fluid present in two adjacent regions of space: in the first region, the fluid is at rest, while in the second it moves at constant velocity. If a small disturbance is introduced at the boundary separating the two regions, fluid particles that were at rest (i.e., with zero velocity) will be moved to the region at finite velocity (and vice versa). This creates an instability: the amplitude of the disturbance continues to increase, and the particles in the two regions will mix together, forming vortices and causing the original configuration to be lost [38–42]. In this simple case, a configuration such as that which has just been described is always unstable, however small the initial disturbance may be [38, 39]. Less rare than might be thought, this phenomenon is often found in nature and can be readily seen with the naked eye. An example are the distinctive wave-like formations that clouds sometimes show if subjected to particular air currents traveling at different speeds.

Kelvin-Helmholtz instability phenomena are thus very frequent on our planet, and studies of their effects have led over the years to the development of models that can predict transitions from laminar to turbulent flow [38, 39]. A basic consequence of this turbulence arising from instability between two fluids is cavitation, which, by causing the gas bubbles that form at the boundary to implode, increases the level of disorder in the fluid [39].

In the last century, many research groups concentrated on cavitation and the effects it has on the surrounding material. The team led by Cardone at the Italian National Research Council conducted experiments with liquid solutions of iron salts cavitated by ultrasounds in which they observed anomalous neutron emissions associated with cavitation [40]. These neutron emissions produced by stable elements such as iron can be explained by assuming that piezonuclear reactions take place: a hypothesis based on an unstable behavior of the material. In solids, brittle fracture is governed by instability phenomena caused both by scale effects and by the material's intrinsic brittleness, whereas in fluids one of the underlying causes of cavitation would appear to be Kelvin-Helmholtz instability.

The Solar System consists of rocky planets – Mercury, Venus, Earth and Mars – with a solid crust, a mantle and a core, and of two planets called "gas giants", Jupiter and Saturn. The latter two planets, composed almost entirely of light elements such as H and He, have a mass that is 317.8 times that of the Earth in the case of Jupiter, and 95.1 times in the case of Saturn. Jupiter orbits the Sun at a distance of 778.33 Million km, and Saturn at a distance of 1429.4 Million km [34]. These two planets at the outer edge of the Solar System have long been the subject of studies in order to achieve a full understanding of their behavior, composition, and origin. Almost all of the probes that have visited Jupiter and Saturn made fly-bys (Pioneer 11, Voyager 1, etc.), or in other words, observed the planets without entering their orbits. The only exceptions were the Galileo probe, which remained in orbit around Jupiter for over 7 years, and the Cassini-Huygens probe, which is now orbiting Saturn [43]. These several missions revealed a number of aspects of Jupiter and Saturn that are of considerable scientific interest. In particular, it was found that both planets emit around twice as much energy as they receive from the Sun [44, 45] and that they are surrounded by a "belt" of radiation 10 times higher than that around the Earth [46, 47].

The "gas giants" Jupiter and Saturn take their name from the fact that their mass is composed almost entirely of gas. In both planets, the enormous masses of gaseous fluid that make up their atmospheres cause gigantic storms that in some cases can last for several centuries. An outstanding example is Jupiter's Great Red Spot, an elliptically-shaped storm consisting of gas and dust seen for the first time more than 300 years ago by Giovanni Cassini [44]. The fluids involved in these huge storms are moved at high velocity, inhomogeneously, giving rise to currents of gas that flow in relative motion. In this way, instability phenomena are created at the boundary between two currents of gas moving at different velocities, as has been observed several times by the many probes that have visited the two planets [43, 44, 48–50]. These flow dynamic instabilities are similar to those observed on the Earth, where they are referred to as Kelvin-Helmholtz instablility, Fig. 14.10.

It was originally thought that both Jupiter and Saturn consisted entirely of light elements such as H and He [48]. However, compositional analyses of the atmospheres of the two planets have shown that, contrary to previous belief, they are also composed of small amounts of heavy elements: NH_3 (0.08 %) and PH_3 (0.0001 %) [48].

Subsequently, a group of scientists focused attention on quantitative variations in these heavy elements on Jupiter [50]. Observing Jupiter's Great Red Spot, they noted that a decrease in the quantity of NH_3 took place around it. They attributed this to convective phenomena that, by transporting this heavy element downwards [50], resulted in a sort of "elimination". The Great Red Spot (GRS) is an enormous storm in Jupiter's southern hemisphere observed for the first time over 300 years ago [44]. The center of the storm is stagnant, and no high-speed gas movements

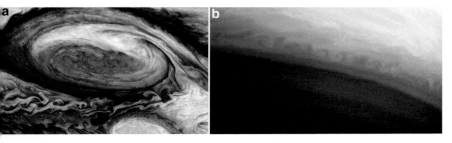

Fig. 14.10 Kelvin-Helmholtz instability around Jupiter's GRS (**a**); Kelvin-Helmholtz instability between currents of gas on Saturn (**b**) [43]

have been detected, while currents of gas and dust at its edges exceed 350 km/h [44]. Using models of the GRS, several scientists have found that this atmospheric phenomenon can produce Kelvin-Helmholtz instability [51]. In this connection, the elimination of heavy elements such as N (deriving from NH_3) at areas affected by Kelvin-Helmholtz instability and thus by possible cavitation, may be caused by piezonuclear reactions involving N, transforming it into lighter elements, e.g. H and He:

$$N_7^{14} \rightarrow H_1^1 + 3He_2^4 + 1 \text{ neutron} \tag{14.7}$$

This could explain the source of the radiation that has always affected the two gas giants but has never been clearly understood [46, 47].

14.5.3   Lithium and Beryllium Depletion on the Sun's Surface

The Sun is the Solar System's mother star, around which the eight major planets (including the Earth) orbit along with the dwarf planets, the satellites, innumerable smaller bodies and the space dust that makes up the interplanetary medium. The Sun's mass, which amounts to around 2×10^{30} kg [52], accounts on its own for 99.8 % of the Solar System's total mass [53, 54].

The energy emitted by the Sun is produced through nuclear fusion processes which compress the nuclei of two or more atoms sufficiently to enable the strong force to overcome electromagnetic repulsion. Consequently, these atoms are fused together to form a single atom, thus generating a nucleus of greater mass than the reacting nuclei as well as, at times, one or more free neutrons. These nuclear fusion reactions take place deep in the Sun's core at temperatures around 13.6×10^6 °K and pressures of 500 Billion atmospheres [49], releasing energy in the form of γ radiation. Once emitted by the core, this radiation is absorbed and re-emitted by the material of the upper layers, contributing to maintain high temperatures; as it travels through the star's layers, the electromagnetic radiation loses energy, assuming longer and longer wavelengths as it passes from the γ band to the x and ultraviolet

Fig. 14.11 Kelvin-
Helmholtz instability in the
Sun's atmosphere [52]

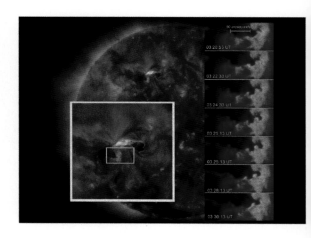

band, reaching the surface at a temperature of around $5507^{\circ\circ}C$ before escaping into space as visible light [55].

The Sun is a medium-small star consisting essentially of hydrogen and helium [56], as well as traces of heavier elements such as C, O, Li and Be [56]. Recent studies of the composition of the Sun's surface have indicated that, contrary to what might be expected from considering these elements as the result of nuclear fusion and from the standard models of a star's evolution, elements such as Li [57–59] and Be [60, 61] are much less abundant than predicted. In evaluating the differences between the present composition of the Sun and that of the proto-Sun, it can be seen that the concentrations of Li are 160 times lower than they were 4.57 Billion years ago [59], which is not what the canonical models predict [62]. Likewise, the concentrations of Be are also lower than predicted. The depletion of these elements is normally associated with convective phenomena that transport them from the surface to hotter areas near the core, where they are destroyed by high temperatures and pressures. However, these convective phenomena are not sufficient to explain such low concentrations of Li and Be on the Sun's surface [58, 60].

It has recently been observed that Kelvin-Helmholtz instability phenomena habitually take place on the Sun's surface and in the so-called "solar corona" [63]. As was indicated earlier, these phenomena may be associated in some way with cavitation and hence with piezonuclear reactions.

It can thus be assumed that the low Li and Be content of the Sun's surface is associated with piezonuclear fission reactions that, through Kelvin-Helmholtz instability, transform Li and Be into lighter elements as follows Fig. 14.11:

$$Be_4^9 \rightarrow 2He_2^4 + 1 \text{ neutron} \tag{14.8}$$

$$Be_4^9 \rightarrow Li_3^6 + H_1^1 + 2 \text{ neutrons} \tag{14.9}$$

$$Li_3^6 \rightarrow He_2^4 + H_1^1 + 1 \text{ neutron} \tag{14.10}$$

14.6   Conclusions

The most important evidence regarding the compositional changes in Mars crust and atmosphere, Mercury crust, Jupiter's Great Red Spot, and Sun's surface may be interpreted in the light of piezonuclear fission reactions. In particular, for the planet Mars, the increment in Fe corresponding to about 2.15×10^{20} kg of Mars' crust, may be counterbalanced by the Ni concentration decrement that is roughly equivalent. These evidence implies Ni as a starting element and Fe as a resultant with the production of H and neutrons (reaction 14.1). In addition, the decrease in K can be directly correlated, through piezonuclear reactions (14.2) and (14.3), to the increase in Cl and an high concentration of an argon isotope, Ar^{36}, in the Mars' atmosphere. From this pint of view, the Mars' atmosphere would appear to be laden with an anomalously high concentration of Ar^{36}. This reaction would appear to be especially interesting as regards the evolution of Mars' proto-atmosphere and its present composition. The correlations between neutron emissions, Fe concentration and fault locations on the planetary crust are surprisingly evident as reported in Fig. 14.7.

Similar evidence may be observed for the planet Mercury. It has a very weak atmosphere consisting of traces of H, He, O [1], and unexpected concentrations of Na [37]. There is still no consensus regarding the phenomenon behind the presence of sodium ions in Mercury's exosphere [37]. An alternative explanation, associated to the planet's tectonic activity, is provided by piezonuclear fission reactions (14.4, 14.5, and 14.6).

In addition, observing Jupiter's Great Red Spot, a decrease in the quantity of NH_3 took place around it. In this connection, the elimination of heavier elements such as N (deriving from NH_3) in areas affected by Kelvin-Helmholtz instability, and thus by turbulence and possible cavitation, may be caused by piezonuclear reaction (14.7), involving N and transforming it into lighter elements, e.g. H and He.

Lastly, it has recently been observed that Kelvin-Helmholtz instability phenomena habitually take place on the Sun's surface and in the so-called "solar corona" [63]. As was indicated earlier, these phenomena may be associated in some way to turbulence. It can thus be assumed that the low Li and Be contents in the Sun's surface could be associated to piezonuclear fission reactions (14.8, 14.9, and 14.10) transforming Li and Be into lighter elements. Piezonuclear reactions are under discussion in the scientific community. Nevertheless the explanation of the Solar System evolution and compositional change provided by the piezonuclear hypothesis seems to have a strong logical basis and multiple cross-checks, mainly as concerned with the justification of very sensitive variations in element concentrations over time.

References

1. Taylor SR, McLennan SM (2009) Planetary crusts: their composition, origin and evolution. Cambridge University Press, Cambridge
2. Hahn BC, McLennan SM (2006) Gamma-ray spectometer elemental abundance correlation with Martian surface age: implication for Martian crustal evolution. Lunar Planet Sci XXXVII:1–2, http://www.lpi.usra.edu/meetings/lpsc2006/pdf/1904.pdf
3. Mitrofanov I et al (2002) Maps of subsurface hydrogen from the high energy neutron detector, Mars Odyssey. Science 297:78–81
4. Knapmeyer M et al (2006) Working models for spatial distribution and level of Mars' seismicity. J Geophys Res 111:E11006. doi:10.1029/2006JE002708
5. Mars Nasa Exploration Sito Web (2011). http://www.nasa.gov/mission_pages/mars/news/mgs-092005.html (consultazione novembre 2011)
6. McSween HY (2007) Mars (chapter 1.22). In: Holland HD, Turekian KK (eds) Treatise on geochemistry update 1. Elsevier Science, Oxford, pp 601–621
7. Righter K, Drake MJ (2000) Metal/silicate equilibrium in the early Earth – new constraints from the volatile moderately siderophile elements Ga, Cu, P and Sn. Geochim Cosmochim Acta 64(20):3581–3597
8. Boynton WV et al (2007) Concentration of H, Si, Cl, K, Fe and Th in the low – and mid – latitude regions of Mars. J Geophys Res 112:E12S99. doi:10.1029/2007JE002887
9. Taylor SR, McLennan SM (1995) The geochemical evolution of the continental crust. Rev Geophys 33:241–265
10. Favero G, Jobstraibizer P (1996) The distribution of aluminum in the Earth: from cosmogenesis to Sial evolution. Coord Chem Rev 149:367–400
11. Jeffrey C et al (2008) Strike-slip faults on Mars: observation and implication for global tectonics and geodynamics. J Geophys Res 113:E08002. doi:10.1029/2007JE002980
12. Carpinteri A, Cardone F, Lacidogna G (2009) Energy emissions from failure phenomena: mechanical, electromagnetic, nuclear. Exp Mech 50:1235–1243
13. Carpinteri A, Cardone F, Lacidogna G (2009) Piezonuclear neutrons from brittle fracture: early results of mechanical compression tests. Strain 45:332–339
14. Cardone F, Carpinteri A, Lacidogna G (2009) Piezonuclear neutrons from fracturing of inert solids. Phys Lett A 373:4158–4163
15. Carpinteri A, Borla O, Lacidogna G, Manuello A (2010) Neutron emissions in brittle rocks during compression tests: monotic vs. cyclic loading. Phys Mesomech 13:264–274
16. Carpinteri A, Chiodoni A, Manuello A, Sandrone R (2011) Compositional and microchemical evidence of piezonuclear fission reactions in rock specimens subjected to compression tests. Strain 47(2):267–281
17. Carpinteri A, Manuello A (2011) Geomechanical and geochemical evidence of piezonuclear fission reactions in the Earth's crust. Strain 47(2):282–292
18. Carpinteri A, Lacidogna G, Manuello A, Borla O (2011) Energy emissions from brittle fracture: neutron measurements and geological evidences of piezonuclear reactions. Strenght Fract Complex 7:13–31
19. Carpinteri A, Lacidogna G, Borla O, Manuello A, Niccolini G (2012) Electromagnetic and neutron emissions from brittle rocks failure: experimental evidence and geological implications. Sadhana 37(1):59–78
20. Carpinteri A, Lacidogna G, Manuello A, Borla O (2012) Piezonuclear fission reactions in rocks: evidences from microchemical analysis, neutron emission, and geological transformation. Rock Mech Rock Eng 45(4):445–459
21. Carpinteri A, Lacidogna G, Manuello A, Borla O (2012) Piezonuclear fission reactions from earthquakes and brittle rocks failure: evidence of neutron emission and nonradioactive product elements. Exp Mech. doi:10.1007/s11340-012-9629-x
22. Carpinteri A, Manuello A (2012) An indirect evidence of piezonuclear fission reactions: geomechanical and geochemical evolution in the Earth's crust. Phys Mesomech 15:14–23

23. Cardone F, Calbucci V, Albertini G (2014) Deformed space-time of the piezonuclear emissions. Mod Phys Lett B 28:1450012
24. Cardone F, Cherubini G, Petrucci A (2009) Piezonuclear neutrons. Phys Lett A 373 (8–9):862–866
25. Widom A, Swain J, Srivastava YN (2015) Photo-disintegratin of the iron nucleus in fractured magnetite rocks with magetostriction. Meccanica 50:1205–1216
26. Widom A, Swain J, Srivastava YN (2013) Neutron production from the fracture of piezoelectric rocks. J Phys G 40:015006. doi:10.1088/0954-3899/40/1/015006
27. Hagelstein PL, Chaubrrdhary IU (2015) Anomalies in fracture experiments, and energy exchange between vibrations and nuclei. Meccanica 50:1189–1203
28. Hartmann WK, Neukum G (2001) Cratering chronology and the evolution of Mars. Space Sci Rev 96:165–194
29. Morgan WJ (1971) Convection plumes in the lower mantle. Nature 230:42–43
30. Van Thienen P, Rivoldini A, Van Hoolst T, Lognonné PH (2006) A top-down origin for mantle plumes. Icarus 185:197–210
31. Wanke H (1991) Chemistry, accretion, and evolution of Mars. Space Sci Rev 56:1–8
32. Carr M (2006) The surface of Mars. Cambridge University Press, New York
33. Faure G, Mensing TM (2007) Introduction to planetary science: the geological perspective. Springer, Dordrecht
34. Rees M (2005) Universe – the definitive visual guide. Dorling Kindersley Ltd, New York. Bouvier A, Wadhwa M (2010) The age of the solar system redefined by the oldest Pb–Pb age of a meteoritic inclusion. Nat Geosci 3:637–641
35. Nittler LR et al (2011) The major-element composition of Mercury's surface from MESSENGER X-ray spectrometry. Science 333:1847–1850
36. Watter TR et al (2009) The tectonics of Mercury: the view after MESSENGER's first flyby. Earth Planet Sci Lett 285:283–296
37. Paral J et al (2010) Sodium ion exosphere of Mercury during MESSENGER flybys. Geophys Res Lett 37:L19102
38. Lord Kelvin (William Thomson) (1871) Hydrokinetic solutions and observations. Philos Mag 42:362–377
39. von Helmholtz H (1868) Über discontinuierliche Flüssigkeits-Bewegungen [On the discontinuous movements of fluids]. Monatsberichte der Königlichen Preussische Akademie der Wissenschaften zu Berlin [Monthly reports of the Royal Prussian Academy of Philosophy in Berlin] 23:215–228
40. Cardone F, Mignani R (2003) Possible observation of transformation of chemical elements in cavitated water. Int J Mod Phys B 17:307–317
41. Franc J-P (2006) Physics and control of cavitation. In: Design and analysis of high speed pumps (pp. 2-1–2-36). Educational Notes RTO-EN-AVT-143, Paper 2. RTO, Neuilly-sur-Seine. Available from: http://www.rto.nato.int/abstracts.asp
42. Aeschlimann V et al (2011) Velocity field analysis in an experimental cavitating mixing layer. Phys Fluids 23:055105. doi:10.1063/1.3592327
43. NASA Exploration Site (2011) http://www.nasa.gov/mission_pages/cassini/main/index.html
44. Rogers JH (1995) The giant planet Jupiter. Cambridge University Press, Cambridge
45. Rieke GH (1975) The thermal radiation of Saturn and its rings. Icarus 1:37–44
46. Brice N et al (1973) Jupiter radiation belt. Icarus 18:206–219
47. Roussos E (2011) Long and short term variability of Saturn's ionic radiation belts. J Geophys Res 116:A02217
48. Lewis JS et al (2004) Physics and chemistry of the solar system. Elsevier Academic press, Amsterdam
49. Irwin GJP (2009) Giant planets of our solar system. Springer, Berlin
50. Fletcher LN et al (2010) Thermal structure and composition of Jupiter's great red spot from high-resolution thermal imaging. Icarus 208:306–328
51. Nezlin MV et al (1982) Kelvin – Helmholtz instability and the Jovian great red spot. JETP Lett 36(6):234–238

52. NASA (2012) http://solarsystem.nasa.gov/planets/profile.cfm?Object=Sun&Display=Facts SunFact Sheet. consultato il 14 Feb 2012
53. Woolfson M (2000) The origin and evolution of the solar system. Astron Geophys 41:1.12–1.19
54. Cohen H (2012) From core to corona: layers of the Sun. Princeton Plasma Physics Laboratory (PPPL). http://fusedweb.llnl.gov/cpep/chart_pages/5.plasmas/sunlayers.html (visited 14 Feb 2012)
55. Hansen CJ et al (2004) Stellar interiors. Springer, New York
56. Basu S, Antia HM (2008) Helioseismology and solar abundances. Phys Rep 457:217–283
57. Blöcker T et al (1998) Lithium depletion in the Sun: a study of mixing based on hydrodynamical simulation. Space Sci Rev 85:105–112
58. Israelian G et al (2009) Enhanced lithium depletion in Sun-like stars with orbiting planets. Nature 462:189–191
59. Baumann P et al (2010) Lithium depletion in solar-like stars: no planet connection. Astron Astrophys. doi:10.1051/0004-6361/201015137
60. Boesgaard AM et al (2002) The case of missing beryllium. http://www.ifa.hawaii.edu/info/press-releases/boes-MissingBeryllium.html
61. Da-run X, Li-cai D (2003) Beryllium depletion in the solar atmosphere. Chin Astron Astrophys 27:1–3
62. D'antona F, Mazzitelli I (1984) Lithium depletion in stars – pre-main sequence burning and extra-mixing. Astron Astrophys 138:431–442
63. Sito Internet NASA (2012) http://www.nasa.gov/mission_pages/sunearth/news/gallery/sun-surf.html. consultato il 15 Feb 2012

Chapter 15
Piezonuclear Fission Reactions Simulated by the Lattice Model of the Atomic Nucleus

Norman D. Cook, Amedeo Manuello, Diego Veneziano, and Alberto Carpinteri

Abstract Recent experiments conducted on natural rocks subjected to different mechanical loading conditions have shown energy emissions in the form of neutrons and anomalous chemical changes. In the present study, a numerical model is used to simulate the nuclear products according to the fission interpretation. Specifically, the reactions were simulated by means of the Lattice Model of the atomic nucleous, assuming nucleons ordered in an antiferromagnetic face-centered-cubic (fcc) array. The simulations indicate that small and middle-sized nuclei can be fractured along weakly-bound planes of the lattice structure. It is argued that the simulations provide theoretical support to the experimentally-observed reactions and, moreover, that the probabilities calculated for various low-energy fissions can be used to explain the stepwise time changes in the element abundances of the Earth's crust, which has evolved from basaltic to sialic composition over geological time.

Keywords Atomic model • Lattice structure • Antiferromagnetic face-centered-cubic (Fcc) array • Atomic nucleus • Low-energy nuclear reactions

15.1 Introduction

In the last few years, numerous experiments have been conducted on natural non-radioactive rocks, such as granite, basalt, magnetite, and marble, by subjecting them to different mechanical loading conditions. The experiments were always accompanied by energy emissions and anomalous chemical changes [1–8], and provided repeatable evidence concerning a new kind of nuclear reaction that may take place during quasi-static or cyclic-fatigue tests at low (2 Hz), intermediate

N.D. Cook (✉)
Department of Informatics, Kansai University, Takatsuki, Osaka 569-1095, Japan
e-mail: cook@res.kutc.kansai-u.ac.jp

A. Manuello • D. Veneziano • A. Carpinteri
Department of Structural, Geotechnical and Building Engineering, Politecnico di Torino, Corso Duca degli Abruzzi 24, 10129 Torino, Italy

© Springer International Publishing Switzerland 2015
A. Carpinteri et al. (eds.), *Acoustic, Electromagnetic, Neutron Emissions from Fracture and Earthquakes*, DOI 10.1007/978-3-319-16955-2_15

(200 Hz), and high (20 kHz) loading frequencies [8]. Such evidence indicates that high-frequency pressure waves, suitably exerted on an inert medium of stable nuclides, can generate neutron emissions and piezonuclear fission reactions [1–14].

Very recently, theoretical interpretations have been proposed by Widom et al. in order to explain neutron emissions as a consequence of nuclear reactions taking place in iron-rich rocks during brittle micro-cracking and fracture [13, 14]. Several evidences show that iron nuclear disintegrations are observed when rocks containing such nuclei are crushed and fractured. The resulting nuclear trasmutations are particularly evident in the case of magnetite rocks and iron-rich materials in general. The same authors argued that neutron emissions may be related to piezoelectric effects and that the fission of iron may be a consequence of the photodisintegration of the same nuclei [13].

The experimental results together with the evidence of the so-called low energy nuclear reactions (LENR) [15–17] strongly suggest that the knowledge of nuclear structure is not a "closed chapter" in Physics. Moreover, recent confirmation of piezonuclear fission reactions, occurring in the Earth's crust and triggered by earthquakes and brittle rocks failure, indicates that old questions concerning nuclear structure should be addressed once again in the light of new phenomena [12]. Even small deviations from conventional assumptions, e.g., concerning the condensation density of nuclear matter or the concept of an average binding energy per nucleon [15, 18–22], could have significant implications. Based on the experimental evidence concerning piezonuclear fission, it would suffice to assume that a nuclear structure failure occurs along weak lattice planes, similar to the cleavage fracture occurring in very hard and strong rocks [6–8].

The nuclear lattice model has been advocated by Cook and Dallacasa as a unification of the diverse models used in nuclear structure theory [17–19, 23–30], but, remarkably, the basic lattice structure was firstly proposed by the originator of the well-established independent-particle model, Eugene Wigner, in 1937 [31] – work that was explicitly cited in his Nobel Prize citation. The lattice model has previously been used to simulate (i) the mass of fission fragments produced by thermal fission of the actinides, and (ii) the transmutation products found on palladium cathodes after electrolysis, as reported in various experimental studies [20, 29, 30, 32]. With regard to the underlying nuclear lattice model, the antiferromagnetic face-centered-cubic (fcc) lattice with alternating proton and neutron layers, is the most suitable model for various reasons: (i) from theoretical research on nuclear matter, it is known to be the lowest-energy solid-phase packing scheme of nucleons $(N = Z)$ [17–19, 22]; (ii) the lattice structure reproduces the quantum number symmetries of the IPM, while being based on the local interactions of the liquid drop model (LDM) [17, 21, 22]; (iii) because of the identity between the nuclear lattice and the IPM, the approximate nucleon build-up procedure is known and implies a specific 3D structure for any given number of protons and neutrons with known quantum numbers, which can be represented in Cartesian space [17].

Several decades of development of the lattice model suggest that the gaseous, liquid, and cluster-phase models of conventional nuclear structure theory can be unified within this specific lattice model. Moreover, the lattice lends itself to straightforward application in explaining different fission modes [17, 21, 22]. The earlier simulation results were concerned with fission fragments from uranium nuclei and transmutation products from palladium isotopes (experimentally reported by Mizuno in 1998 and 2000 [20, 32]). In the present simulations, the Nuclear Visualization Software (NVS) [17] was used to simulate the anomalous nuclear reactions recently observed by Carpinteri et al. [1–11] and to numerically reproduce the nuclear products observed after fracture and fatigue experiments. The results lead to the conclusion that the anomalous nuclear reactions, emerging from Energy Dispersive X-Ray Spectroscopy (EDS) analysis of fractured specimens and from the evolution of the continental Earth's crust, can be well explained by the nuclear lattice model. In addition, the lattice approach allows one to compute the probability related to each possible fission reaction. The probability values obtained for the anomalous reactions can then be used to interpret the evolution of the abundance of product elements in the Earth's crust, oceans, and atmosphere.

15.2   Reproducing Anomalous Fission Fragments Using the Lattice Model

The nuclear lattice model proposed by Cook and Dallacasa [17–19, 23–28] can be used to simulate the piezonuclear fission reactions by constructing individual iso-topes, in accordance with the lattice build-up procedure, and then simulating the cleavage of the lattice along various lattice planes. The starting points for the simulations are therefore the nuclear structures of the elements and those known to be abundant in the Earth's crust today and in previous eras [6–12]. Although other nuclear structure models have been developed since the 1930s, the fcc lattice is the most suitable to simulate the anomalous reactions recently discovered because of its clear structural implications. The simulation begins with a 3D lattice structure of specific isotopes based on the total number of neutrons N and protons Z. By simulating the fission of the nucleus as a fracture occurring along a certain section plane across the lattice, "fragments" are produced, and correspond to the post-fission daughter nuclei.

The quantum mechanical foundations of the lattice model and its relation to the Schrödinger wave-equation have been discussed elsewhere, but, for the purposes of the simulation, it is sufficient to describe the lattice structure in Cartesian space. That is, the mean position of each nucleon can be defined in relation to its quantum numbers by means of the following equations [17]:

$$x = |2m|(-1)(m + 1/2) \tag{15.1}$$

$$y = (2j + 1 - |x|)(-1)^{(i+j+m+1/2)} \tag{15.2}$$

$$z = (2n + 3 - |x| - |y|)(-1)^{(i+n-j-1)} \tag{15.3}$$

where n, m, j, s, i are the quantum numbers that describe the energy state for a given nucleon [21, 22]. Equations 15.1, 15.2, and 15.3 are deduced from a rigorous, self-consistent representation of the IPM in three-dimensional space [17]. In accordance with the known quantum mechanics of nuclear states, each nucleon is characterized by a unique set of five quantum numbers, which define the precise energy state of the nucleon, as described by the Schrödinger equation [17, 21, 22]. The spatial origin of this wave is a function of the three coordinates x, y, z. Hence, for any isotope, knowing that each nucleon belongs to a certain energy level given by the values of its quantum numbers, it is possible to consider the 3D lattice of nucleons as a representation of its quantum mechanical state [17, 21, 22].

Piezonuclear reactions cannot be defined as traditional fission reactions, since temperature and energy conditions are not equivalent to those involved in thermal neutron-induced fission. For this reason, it is convenient to verify that these anomalous reactions may be correctly described in the NVS simulations, as already shown concerning fission fragments from uranium and transmutation products from palladium [20, 29, 30, 32]. To simulate this new kind of fission, we assumed that fractures occur in the nuclear lattice along their crystal planes. Following this approach, two distinct fragments are produced from any considered reaction. Their characteristics are given by the NVS in terms of fragment stability, fission threshold energy along a certain fracture plane, and the number of protons and neutrons in each fragment. The resulting elements can be deduced from the characteristics of the fragments obtained at the end of the simulation. The analysis of the nuclear characteristics is described in the next section along with the isotopes obtained from the piezonuclear reactions. Important considerations are made on the neutron emissions from the anomalous nuclear reactions measured during the experiments. They may be deduced by investigating the stability of the resulting fragments. We assume that unstable isotopes with an excess of neutrons are likely to induce neutron emissions, depending on local binding characteristics, in order to reach more stable nuclear states. These emissions deduced from the model are then compared with the experimental results reported by Carpinteri et al. [1–11].

15.3 Simulations and Results

The NVS simulates lattice structures up to 480 nucleons and calculates fission results along 17 different section planes for each given nuclide (Table 15.1). In fact, many more lattice planes are available for simulation, but the electrostatic repulsion between the protons in the two fragments is much greater for lattice planes that break the lattice structure into approximately symmetrical fragments, so that many low-repulsion, asymmetrical fission events are ignored. A single simulation consists in fracturing the nucleus along one single plane at a time, breaking only the bonds that connect the two fragments [17]. It is understandable that the choice of the fracture planes is affected by the lattice structure and, therefore, by the position of the nucleons with respect to the x, y, z axes. Being a lattice model drawn from

Table 15.1 Fission planes and their identification number in the NVS

Fracture plane	Equation
1	$x = 2$
2	$x = 0$
3	$x = -2$
4	$z = -2$
5	$z = 0$
6	$z = 2$
7	$y = 2$
8	$y = 0$
9	$y = -2$
10	$-x + y + z + 1 = 0$
11	$-x + y + z - 1 = 0$
12	$x - y + z + 1 = 0$
13	$x - y + z - 3 = 0$
14	$-x - y + z - 1 = 0$
15	$-x - y + z + 3 = 0$
16	$x + y + z - 1 = 0$
17	$x + y + z + 3 = 0$

crystallography implies that all the planes used for the simulations correspond to the principal crystallographic planes [17]. In particular, the 17 planes used by the NVS are parallel to the horizontal, vertical and inclined (45°) planes passing through or near the origin of the axes. Each plane is identified by a number from 1 to 17, as shown in Table 15.1.

Eighteen reactions derived from direct and indirect experimental evidence were simulated by means of the NVS along different fission planes and each considering a different starting element (Table 15.2). As mentioned in the Introduction, the elements known to be involved in the piezonuclear reactions were considered in the numerical simulations. Such reactions are strictly connected to: (i) the experimental results obtained from the EDS analyses performed after fracture tests on natural rock specimens, or (ii) the compositional changes in the Earth's crust evolution during the last 4.57 billion years [5–12]. As recently reported [9–12], the evolution of the Earth's crust and atmosphere, the formation of oceans and greenhouse gases, and the origin of life are phenomena deeply related to piezonuclear reactions [1–12]. This was the motivation for undertaking the simulation of fission reactions according to a non-traditional methodology [19].

These reactions were simulated using two build-up procedures for nuclear structure. The first one generates a default nucleus where each nucleon has a pre-assigned position and quantum numbers that give the lattice a regular, densely-packed, polyhedral structure. The second procedure uses the "picking function", which is the most convenient way for constructing a nucleus from a set of nucleons with specific quantum numbers and coordinates [17]. The picking function was applied only when the lattice structures using the default configuration were inappropriate due to weak bonding of the last few nucleaons. In particular,

Table 15.2 Piezonuclear reactions obtained as direct evidence from EDS analysis of fractured specimens, or conjectured considering the continental Earth's crust evolution

Earth's crust evolution	
(15.1)	$Fe_{26}^{56} \rightarrow 2\,Al_{13}^{27} + 2$ neutrons
(15.2)	$Fe_{26}^{56} \rightarrow Mg_{12}^{24} + Si_{14}^{28} + 4$ neutrons
(15.3)	$Fe_{26}^{56} \rightarrow Ca_{20}^{40} + C_{6}^{12} + 4$ neutrons
(15.4)	$Co_{27}^{59} \rightarrow Al_{13}^{27} + Si_{14}^{28} + 4$ neutrons
(15.5)	$Ni_{28}^{59} \rightarrow 2\,Si_{14}^{28} + 3$ neutrons
(15.6)	$Ni_{28}^{59} \rightarrow Na_{11}^{23} + Cl_{17}^{35} + 1$ neutron
Atmosphere evolution, Ocean formation and origin of life	
(15.7)	$Mg_{12}^{24} \rightarrow 2C_{6}^{12}$
(15.8)	$Mg_{12}^{24} \rightarrow Na_{11}^{23} + H_{1}^{1}$
(15.9)	$Mg_{12}^{24} \rightarrow O_{8}^{16} + 4H_{1}^{1} + 4$ neutrons
(15.10)	$Ca_{20}^{40} \rightarrow 3C_{6}^{12} + He_{2}^{4}$
(15.11)	$Ca_{20}^{40} \rightarrow K_{19}^{39} + H_{1}^{1}$
(15.12)	$Ca_{20}^{40} \rightarrow 2O_{8}^{16} + 4H_{1}^{1} + 4$ neutrons
Greenhouse gas formation	
(15.13)	$O_{8}^{16} \rightarrow C_{6}^{12} + He_{2}^{4}$
(15.14)	$Al_{13}^{27} \rightarrow C_{6}^{12} + N_{7}^{14} + 1$ neutron
(15.15)	$Si_{14}^{28} \rightarrow 2\,N_{7}^{14}$
(15.16)	$Si_{14}^{28} \rightarrow C_{6}^{12} + O_{8}^{16}$
(15.17)	$Si_{14}^{28} \rightarrow 2C_{6}^{12} + He_{2}^{4}$
(15.18)	$Si_{14}^{28} \rightarrow O_{8}^{16} + 2He_{2}^{4} + 2H_{1}^{1} + 2$ neutrons

once a specific default nucleus has been constructed and displayed, usually the last two protons or neutrons were individually moved from one energy-state to another to find the configuration that best reproduces the product elements.

All 18 reactions reported in Table 15.2 are simulated by the NVS and the results are summarized in Table 15.3. For each simulation, the plane that allows the anomalous fission is indicated. Each fragment is identified by the number of protons (Z), the number of neutrons (N), and the corresponding isotope. In addition, when the fragment is unstable, NVS displays the experimentally-known decay time of that fragment. In the case of unstable fission fragments, the number of neutrons exceeds the stable condition and neutron emissions may occur from that fragment in order to achieve stability. It is interesting to note that, from the NVS results, it is possible to reproduce the neutron emissions of piezonuclear reactions, in addition to the product elements (fragments).

For the first simulation, the Fe_{26}^{56} nucleus was chosen as the starting element and the lattice of this nucleus is shown in Fig. 15.1. Its characteristics assigned by the software are as follows: Protons: 26; Neutrons: 30; n-values: 0, 1, 2, 3; j-values: 1/2, 3/2, 5/2, 7/2; m-values: $\pm 1/2$, $\pm 3/2$, $\pm 5/2$, $\pm 7/2$, according to the literature [17, 21, 22].

Table 15.3 Summary of the fragment characteristics obtained from simulations conducted according to the 18 piezonuclear reactions: the neutron emission column represents the sum of the neutrons released from both fragments for each fission plane

Reaction	Plane	Fragment 1					Fragment 2					Neutron Emission	Fission Probability (%)
		Z	N	A	Isotope	Decay time[a]	Z	N	A	Isotope	Decay time		
(1)[a]	2	13	14	27	Al^{27}	Stable	13	16	29	Al^{27}	6.6 [m]	+2n	23.22
	8	13	14	27	Al^{27}	Stable	13	16	29	Al^{27}	6.6 [m]	+2n	
(2)[b]	14	12	12	24	Mg^{24}	Stable	14	18	32	Si^{28}	172[y]	+4n	40.74
	16	12	12	24	Mg^{24}	Stable	14	18	32	Si^{28}	172[y]	+4n	
(3)[b]	1	6	7	13	C^{12}	Stable	20	23	43	Ca^{40}	Stable	+4n	36.04
	3	20	22	42	Ca^{40}	Stable	6	8	14	C^{12}	5730[y]	+4n	
(4)	2	13	14	27	Al^{27}	Stable	14	18	32	Si^{28}	172 [y]	+4n	100
	8	14	16	30	Si^{28}	Stable	13	16	29	Al^{27}	6.6 [m]	+4n	
(5)[b]	2	14	16	30	Si^{28}	Stable	14	15	29	Si^{28}	Stable	+3n	32.60
(6)[b]	12	11	13	24	Na^{23}	14.95[h]	17	18	35	Cl^{35}	Stable	+n	67.40
(7)[b]	2	6	6	12	C^{12}	Stable	6	6	12	C^{12}	Stable		11.26
(8)[a]	9	11	12	23	Na^{23}	Stable	1	0	1	H^{1}	Stable		31.01
(9)[b]	4	8	12	20	O^{16}	13.5[s]	4	0	4	4H^{1}	Stable	+4n	57.73
	6	0	4	4	4H^{1}		12	8	20	O^{16} + 4H^{1}	0.1[s]	+4n	
	7	4	4	8	4H^{1}	0.07[fs]	8	8	16	O^{16}	Stable	+4n	
	8	8	9	17	O^{16}	Stable	4	3	7	4H^{1}	0.07[fs]	+4n	
(10)[b]	1	2	2	4	He^{4}	Stable	18	18	36	3C^{12}	Stable	+4n	10.28
(11)[b]	1	1	0	1	H^{1}	Stable	19	20	39	K^{39}	Stable		58.46
(12)	1	4	4	8	4H^{1}	0.07[fs]	16	16	32	2O^{16}	Stable	+4n	31.26
	3	16	16	32	2O^{16}	Stable	4	4	8	4H^{1}	0.07[fs]	+4n	
	7	4	4	8	4H^{1}	0.07[fs]	16	16	32	2O^{16}	Stable	+4n	
	9	16	16	32	2O^{16}	Stable	4	4	8	4H^{1}	0.07[fs]	+4n	

(continued)

Table 15.3 (continued)

Reaction	Plane	Fragment 1					Fragment 2					Neutron Emission	Fission Probability (%)
		Z	N	A	Isotope	Decay time[a]	Z	N	A	Isotope	Decay time		
(13)[b]	13	6	6	12	C^{12}	Stable	2	2	4	He^4	Stable		100
(14)	2	6	7	13	C^{12}	Stable	7	7	14	N^{14}	Stable	+n	100
(15)	2	7	7	14	N^{14}	Stable	7	7	14	N^{14}	Stable		14.66
	8	7	7	14	N^{14}	Stable	7	7	14	N^{14}	Stable		
(16)	10	6	6	12	C^{12}	Stable	8	8	16	O^{16}	Stable		35.89
	12	6	6	12	C^{12}	Stable	8	8	16	O^{16}	Stable		
	14	6	6	12	C^{12}	Stable	8	8	16	O^{16}	Stable		
	16	6	6	12	C^{12}	Stable	8	8	16	O^{16}	Stable		
(17)	9	12	12	24	$2\,C^{12}$	Stable	2	2	4	He^4	Stable		15.06
(18)	1	2	3	5	He^4	0.76[zs]	12	11	23	$O^{16}+He^4+2H^1$	11.3[s]	+2n	34.38
	8	8	9	17	O^{16}	Stable	6	5	11	$2He^4+2H^1$	20.385[m]	+2n	
	13	12	11	23	$O^{16}+He^4 + 2H^1$	11.3[s]	2	3	5	He^4	0.76[zs]	+2n	

[a] Decay time of the isotope (half life): zepto-seconds, 10^{-21} (zs); femto-seconds, 10^{-15} (fs); seconds (s); minutes (m); hours (h); days (d); years (y)
[b] Picking option

Fig. 15.1 Lattice structure
of Fe^{56}

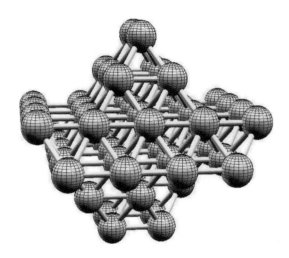

Once the nuclear structure is modified with the picking option, the lattice is "fractured" along fission planes that cut all nucleon-nucleon bonds connecting the two fragments (Fig. 15.2). Among the 17 planes of fission, six of them are relevant for reactions (15.1), (15.2), and (15.3) of Table 15.2, whereas the remaining 11 produce results unrelated to the empirical data (Table 15.2). Specifically, the simulation results related to reaction (15.1) occur along fission planes 2 and 8, which are the yz and the xz plane in Cartesian space respectively, see Table 15.1 and Fig. 15.2. As shown in Table 15.4, the fragments produced from fissions along planes 2 and 8 possess the same characteristics as those described in the piezonuclear reaction (15.1) (Table 15.2). The two fragments correspond to Al^{27}_{13}, and Al^{29}_{13}. The former is stable, whereas the latter is unstable as it contains two neutrons in excess, which are weakly-bound to the lattice fragment and presumably emitted when the reaction occurs. Assuming the emission of these two neutrons, the fission can be considered as symmetric with respect to both planes 2 and 8.

The second simulation was run in accordance with reaction (15.2) (see Tables 15.2 and 15.3). In this case, the lattice structure for Fe is produced using the picking option [17]. The fragments obtained from the fission simulated by NVS are summarized in Table 15.5. The results of the simulation show that the fragments are consistent with those of reaction (15.2) for fissions occurring along two different planes: 14, 16 (see Table 15.5 and Fig. 15.3). It is remarkable that the fragments from the fission along plane 14 are identical to those along plane 16. The simulation produces an isotope of Mg and an isotope of Si, which are identified as Fragment 1 and Fragment 2. In each case, Fragment 1 is stable, whereas Fragment 2 is unstable. In particular, Fragment 1 is a stable nucleus of Mg^{24} and Fragment 2 is a nucleus of Si^{32}, unstable, which contains 4 neutrons that can be emitted when reaction (15.2) occurs.

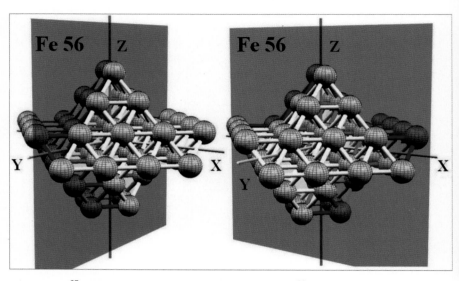

Fig. 15.2 Al^{27} fragments of reaction (15.1): fracture of the Fe^{56} lattice along planes 2 and 8

Table 15.4 Fragments $\left(2\,Al_{13}^{27} + 2\ \text{neutrons}\right)$ from the simulation of reaction (15.1)

Fission plane	Fragment 1				Fragment 2					Partial fission probability (%)
	Z_1	N_1	A_1	Isotope	Z_2	N_2	A_2	Isotope		
2	13	14	27	Al^{27}	13	16	29	Al^{27}	+2n	10.65
8	13	14	27	Al^{27}	13	16	29	Al^{27}	+2n	12.56

Table 15.5 Fragments $\left(Mg_{12}^{24} + Si_{14}^{28} + 4\ \text{neutrons}\right)$ from the simulation of reaction (15.2)

Fission plane	Fragment 1				Fragment 2					Partial fission probability (%)
	Z_1	N_1	A_1	Isotope	Z_2	N_2	A_2	Isotope		
14	12	12	24	Mg^{24}	14	18	32	Si^{28}	+4n	20.37
16	12	12	24	Mg^{24}	14	18	32	Si^{28}	+4n	20.37

Considering the data from the Earth's crust and the indirect piezonuclear evidence, reactions (15.7) and (15.12) are particularly significant for their implications concerning atmosphere evolution and ocean formation, respectively [9–12]. The simulation results of these reactions are summarized in Table 15.3. The simulation of piezonuclear reaction (15.7) produces as fragments two nuclei of C^{12} with no excess of neutrons (Figs. 15.4 and 15.5). Therefore, the simulation describes a symmetrical fission that does not entail neutron emissions according to piezonuclear reaction (15.7).

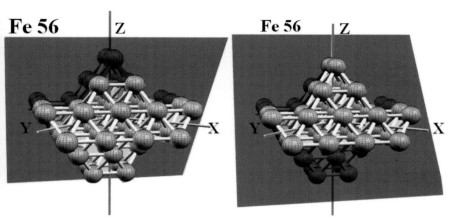

Fig. 15.3 Mg^{24} and Si^{28} fragments from reaction (15.2): fracture of Fe^{56} lattice along planes 14 and 16

Fig. 15.4 Mg^{24} nuclear lattice structure

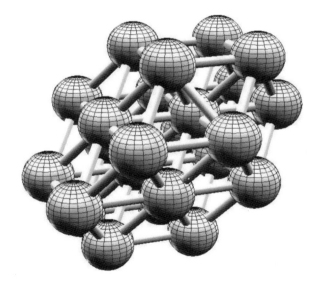

On the other hand, as observed in the case of reactions (15.1), (15.2), and (15.12), there are unstable fragments. In particular, for every plane (Table 15.3) that allows for the products $\left(2O_8^{16} + 4H_1^1\right)$ of piezonuclear reaction (15.12), at least one of the two fragments obtained from each simulation is unstable. This result suggests that neutron emission is favored in many of the lattice fission events (Figs. 15.6 and 15.7).

Fig. 15.5 Fragments from reaction (15.7): fracture of the Mg^{24} lattice along plane 2

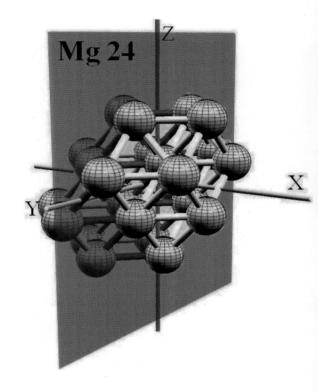

Fig. 15.6 Ca^{40} nuclear lattice structure

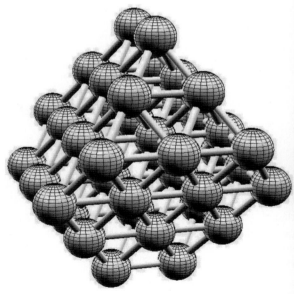

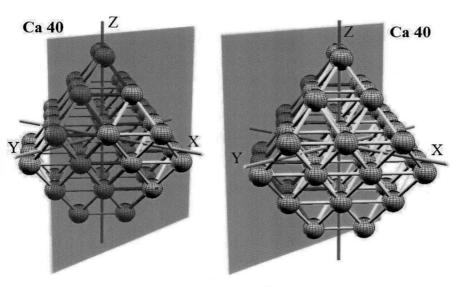

Fig. 15.7 Fragments of reaction (15.12): fracture of Ca^{40} along planes 1 and 3

15.4 Binding Energy and Probability of Alternative Piezonuclear Fission Reactions

As described above, the nucleus is represented as a lattice where the nodal positions are occupied by the nucleons. A given nuclear lattice in its ground-state has a certain total binding energy (BE) that generally depends on the number and type of nearest-neighbor nucleon-nucleon bonds, thus on the number of constituent nucleons [21, 22]. The binding energy is usually expressed as average binding energy per nucleon (BE/nucleon) or average binding energy per bond (BE/bond), which represents a mean value of the energy distribution among the bonds in the nucleus [21, 22]. According to the lattice model, the bonds are not all equivalent and are formed by various combinations of nucleon states (as specified by quantum numbers n, j, m, s and i). Specifically, the dipole-dipole interactions of nucleon pairs are attractive (singlet-pairs) for all nearest-neighbor PP and NN combinations, but there are both attractive and repulsive dipole combinations for PN pairs (triplet- and singlet-pairs, respectively). As a consequence, the antiferromagnetic fcc lattice with alternating proton-neutron layers implies the existence of lattice planes that are either strongly or weakly bound, depending on the character of the bonds in the lattice plane. It is the internal structure of the nucleon lattice that leads directly to the prediction of lattice fragments of various masses and probabilities.

In order to assess the probability of fission occurring in a given nucleus, it is necessary to know the binding force of the nuclear lattice structure: the lower is this force the higher is the probability of fission. This can be evaluated in terms of binding energy of the lattice structure through a specific plane. In particular, the binding energy of the bonds between nearest-neighbor nucleons crossed by a fracture plane minus the Coulomb repulsion through the same plane represents the residual binding energy. Inverting this value yields a ratio (MeV^{-1}) that is defined as proportional to the probability of fission ($P_{fission}$) [17]:

$$P_{fission}(Z,N) = \frac{1}{\left(\beta \sum_{m}^{A_{f1}} \sum_{m}^{A_{f2}} b_{m,n} - \sum_{j}^{Z_{f1}} \sum_{k}^{Z_{f2}} Q_{j,k} \right)} \qquad (15.4)$$

where Z, N, and A are the number of protons, neutrons and the total number of nucleons contained in the atomic nucleus; f_1 and f_2 stand for the two resulting fragments of the given reaction; β is an experimental value of the nucleon-nucleon binding force. Between nearest-neighbor nucleons this value changes according to the different nature of the bond; $b_{m,n}$ is the number of bonds across the fission plane taken into consideration and so the number of broken nucleon-nucleon bonds along the fracture plane; $Q_{j,k}$ is the Coulomb repulsive contribution between the protons in the two fragments defined by the fission plane.

Using the parameter $P_{fission}$, the probabilities of piezonuclear reactions (15.1) and (15.2) were calculated as shown in detail in Tables 15.4 and 15.5 and summarized in Table 15.3. The Fission Probability is expressed as a normalized percentage of the cases studied for each element reported in Table 15.2 with their relevant fission planes, and a given binding force value β is assumed (approximately comprised in the range between 2 and 4 MeV) [17]. From Table 15.4, the simulation of piezonuclear reaction (15.1) results in two cases of the 17 possible fission planes, with a total $P_{fission}$ of ~23 %, whereas from Table 15.5 it is observed that reaction (15.2) results in two cases having a total $P_{fission}$ of ~41 %. According to these considerations, it is of interest that these probabilities reproduce the known abundances of the Al, Mg and Si elements in the Earth's crust [9–12]. The probability that a nucleus of Fe produces Magnesium and Silicon (piezonuclear reaction (15.2) in Table 15.2) is significantly larger (ratio:1.74) than that implying the symmetrical nuclear fission of Fe into two Al atoms (reaction (15.1) in Table 15.2). This is in agreement with the evidence regarding the compositional changes in the evolution of the Earth's crust. In fact, the total decrease in Fe over the last 4.57 Billion years of about 11 % seems to be consistently counterbalanced by the increases in Mg, Si and Al, where the contribution of Mg and Si ~7 % is a little less than twice that of the Al increase, ~4 % [9–12]. The ratio of the normalized $P_{fission}$ of reaction (15.2) to that of reaction (15.1) is approximately 1.75, as the ratio of Mg and Si (~7 %) increase to the increase in Al (~4 %) in the Earth's crust. It is also interesting to note that reaction (15.3) involving Fe as the starting element and Ca and C as the resultants can be obtained by the NVS simulation. This reaction, not so frequent in the Earth's crust system, could be recognized as a fundamental reaction during the application of ultrasound to sintered Ferrite (α-Iron) and steel bars, as recently reported by Cardone et al. [33]. This evidence indicates that the NVS is able to reproduce different piezonuclear fission reactions also belonging to different systems and experiments at different scales (Earth's crust or ferrite bar).

The results obtained from the Ca-based reaction simulations also showed consistency with the findings reported in [9–12]. During the evolution of the Earth's crust, Ca decreased by about ~4 %, while K increased by about ~2.7 % according to reaction (15.11). The total Ca depletion may be almost perfectly counterbalanced considering the increases in O and H (H_2O) that together correspond to an increment of about 1.3 % [9–12]. This means that more than two thirds of the Ca depletion resulted in potassium and approximately one third in H_2O. On the other hand, considering the probability of fission computed for reactions (15.10), (15.11) and (15.12) by NVS (Table 15.3), the piezonuclear reaction involving K as the product, reaction (15.11), returned a normalized probability of ~58.4 %. The normalized probability of the simulation of reaction (15.12), involving H and O as products, is about 31.2 % (see Table 15.3) [9–12]. The ratio of the fission probability of reaction (15.11) to that of reaction (15.12) is about two, which is in good agreement with the evidence concerning the Earth's crust [9–12].

The consistency with the evidence drawn from the Earth's crust can be verified also in the case of Mg as the starting element of the anomalous reactions (15.7–9). In particular, we find a global Mg decrease (~7.9 %) that is counterbalanced by a ~2.7 % increase in Na in the Earth's crust and by increases of ~2.0 % and ~3.2 % in H_2O and C, respectively, in the ancient atmosphere. The evidence of the Na increase is supported by the NVS simulation, that returned a normalized probability percentage for reaction (15.8) equal to ~31.0 %. With regard to the other reactions, involving C, O and H as resultants, we obtained a total normalized probability percentage of about ~69 %. These last two percentages are in good agreement with the considerations concerning the evolution of the Earth's crust composition as about two thirds of the Mg decrease can be ascribed to the formation of gaseous elements such as C and H_2O that formed in the proto-atmosphere of our planet [9–11, 34].

15.5 Conclusions

The simulations conducted using the NVS reproduced the piezonuclear reactions observed in both laboratory tests and Earth's crust by Carpinteri et al. [1–12]. The results were obtained using the approach recently proposed by Cook and co-workers to provide a fully quantum mechanical unification of the different models of nuclear structure [17, 20].

The issue still remains of how strain energy would be coupled to nuclei as a crack propagates sufficiently strongly and efficiently so that nuclei would sometimes split rather than the crack going between atoms and leaving nuclei intact. Precisely such a mechanism for piezoelectric and piezomagnetic rocks is discussed quantitatively in references [13, 14] and the nuclear lattice model here offers a way to discuss the expected fission products beyond the simple symmetrical fission of a liquid drop suggested by Swain et al. [14]. For non-piezoelectric and non-piezomagnetic rocks, a similar effect has been proposed by Preparata [35].

The simulations suggest that neutron emissions can be favored when the product fragments present unstable conditions. From this point of view, the recent evidence provided by fracture and fatigue experiments indicates neutron emissions far in excess of the background level [1–12]. This may be correlated to piezonuclear fission of nuclei along specific weak planes of the lattice structure. Furthermore, the presence of relatively weakly-bound planes within the lattice can be assumed as an indicator of the lattice behavior, through a given plane, calculated by means of the Fission Probability. This implies that certain piezonuclear reactions may occur with a higher probability than others. In particular, the total decrement in Fe, over the last 4.57 billion years, of about 11 % consistently counterbalanced by an increase in Mg and Si (~7 %), and by that in Al (~4 %), is confirmed by the normalized fission probabilities of the relevant reactions computed by the lattice model. Analogously, similar numerical results obtained by NVS supported the decrement in Ca and Mg and the increments in K, Na, C and H2O, contributing to explain the Earth's crust evolution together with the proto-atmosphere composition and the formation of the oceans, under the light of the piezonuclear conjecture.

Finally, the current version of the NVS simulates the fission of nuclei assuming an average value of the nuclear binding force uniformly distributed. More precise results could be obtained in the case of these reactions using an improved version of the NVS that is able to consider the binding energy between nucleons as a function of the different nucleon states. Such improvements will be implemented in future research in order to take into consideration a more realistic distribution of the binding energy across the nucleus.

References

1. Carpinteri A, Cardone F, Lacidogna G (2009) Piezonuclear neutrons from brittle fracture: early results of mechanical compression tests. Strain 45:332–339. Atti dell'Accademia delle Scienze di Torino 33:27–42
2. Cardone F, Carpinteri A, Lacidogna G (2009) Piezonuclear neutrons from fracturing of inert solids. Phys Lett A 373:4158–4163
3. Carpinteri A, Cardone F, Lacidogna G (2010) Energy emissions from failure phenomena: mechanical, electromagnetic, nuclear. Exp Mech 50:1235–1243
4. Carpinteri A, Borla O, Lacidogna G, Manuello A (2010) Neutron emissions in brittle rocks during compression tests: monotonic vs. cyclic loading. Phy Mesomech 13:268–274
5. Carpinteri A, Lacidogna G, Manuello A, Borla O (2011) Energy emissions from brittle fracture: neutron measurements and geological evidences of piezonuclear reactions. Strenght Fract Complex 7:13–31
6. Carpinteri A, Lacidogna G, Manuello A, Borla O (2012) Piezonuclear fission reactions: evidences from microchemical analysis, neutron emission, and geological transformation. Rock Mech Rock Eng 45:445–459
7. Carpinteri A, Lacidogna G, Borla O, Manuello A, Niccolini G (2012) Electromagnetic and neutron emissions from brittle rocks failure: experimental evidence and geological implications. Sadhana 37:59–78

8. Carpinteri A, Lacidogna G, Manuello A, Borla O (2013) Piezonuclear fission reactions from earthquakes and brittle rocks failure: evidence of neutron emission and nonradioactive product elements. Exp Mech 53(3):345–365
9. Carpinteri A, Chiodoni A, Manuello A, Sandrone R (2011) Compositional and microchemical evidence of piezonuclear fission reactions in rock specimens subjected to compression tests. Strain 47(2):267–281
10. Carpinteri A, Manuello A (2011) Geomechanical and geochemical evidence of piezonuclear fission reactions in the Earth's crust. Strain 47:282–292
11. Carpinteri A, Manuello A (2012) An indirect evidence of piezonuclear fission reactions: geomechanical and geochemical evolution in the Earth's crust. Phys Mesomech 15:14–23
12. Manuello A, Sandrone R, Guastella S, Borla O, Lacidogna G, Carpinteri A (2016) Neutron emissions and compositional changes at the compression failure of iron-rich natural rocks. In: Carpinteri A, Lacidogna G, Manuello A (eds) Acoustic, electromagnetic, neutron emissions from fracture and earthquakes. Springer, New York
13. Widom A, Swain J, Srivastava YN (2015) Photo-disintegratin of the iron nucleus in fractured magnetite rocks with magetostriction. Meccanica 50(5):1205–1216
14. Widom A, Swain J, Srivastava YN (2013) Neutron production from the fracture of piezoelectric rocks. J Phys G 40:015006
15. Bridgman PW (1927) The breakdown of atoms at high pressures. Phys Rev 29:188–191
16. Storms E (2007) Science of low energy nuclear reaction: a comprehensive compilation of evidence and explanations about cold fusion. World Scientific Publishing Co, Singapore
17. Cook ND (2010) Models of the atomic nucleus, 2nd edn. Springer, Berlin
18. Cook ND, Dallacasa V (1987) Face-centered-cubic solid-phase theory of the nucleus. Phys Rev C 35:1883–1890
19. Cook ND, Dallacasa V (2012) LENR and nuclear structure theory. In: Proceedings of ICCF-17, Daejong, 12–17 Aug
20. Mizuno T (1998) Nuclear transmutation: the reality of cold fusion. Tuttle, Concord
21. Lilley J (2001) Nuclear physics: principles and applications. Wiley, Chichester
22. Krane KS (1987) Introductory nuclear physics. Wiley, Chichester
23. Cook ND (1976) An FCC lattice model for nuclei. Atomkernenergie 28:195–199
24. Cook ND (1981) Quantization of the FCC nuclear theory. Atomkernenergie 40:51–55
25. Dallacasa V, Cook ND (1987) The FCC nuclear model (II). Il Nuovo Cimento 97:157–183
26. Cook ND (1989) Computing nuclear properties in the FCC model. Comput Phys 3:73–77
27. Cook ND (1994) Nuclear binding energies in lattice models. J Phys G 20:1907–1917
28. Cook ND (1999) Is the lattice gas model a unified model of nuclear structure? J Phys 25:1213–1221
29. Cook ND (2008) Toward an explanation of transmutation products on palladium cathodes. In: Proceedings of ICCF-14, Rome
30. Cook ND (2010) Simulation of palladium fission products using the FCC lattice model. In: Proceedings of ICCF-16, Chennai
31. Wigner E (1937) On the consequences of the symmetry of the nuclear Hamiltonian on the spectroscopy of nuclei. Phys Rev 51:106–119
32. Mizuno T (2000) Experimental confirmation of the nuclear reaction at low energy caused by electrolysis in the electrolyte. In: Proceedings of the symposium on Advanced Research in Energy Technologh, Hokkaido University, Hokkaido. March 15–17, pp 95–106
33. Cardone F, Manuello A, Mignani R, Petrucci A, Santoro E, Sepielli M, Carpinteri A (2016) Ultrasonic piezonuclear reactions in steel and sintered ferrite bars. J Adv Phy 5:1–7
34. Liu L (2004) The inception of the oceans and CO_2-athmosphere in the early history of the Earth. Earth Planet Sci Lett 227:179–184
35. Preparata GA (1991) New look at solid-state fractures, particle emission and "Cold" nuclear fusion. Il Nuovo Cimento 104(8):1259–1263

Chapter 16
Correlated Fracture Precursors in Rocks and Cement-Based Materials Under Stress

Gianni Niccolini, Oscar Borla, Giuseppe Lacidogna, and Alberto Carpinteri

Abstract This paper presents experimental results on the evolution of damage by acoustic-emission and electrical resistance measurements in rock and cement mortar specimens during uniaxial compression tests. Once defined a specific damage parameter in terms of cumulated number of acoustic emission events, evaluated by their magnitude, two scaling laws are proposed which correlate respectively the electrical resistance variation and the acoustic emission b-value with the cumulative damage D. The electrical resistance variation is expressed as the ratio R_0/R, where R_0 is the resistance of the undamaged specimen and R is that obtained during the test. The first scaling law describes a relevant correlation between acoustic emission and electrical resistance measurements, while the second one shows internal consistency of two metrics both derived from acoustic emission data.

Keywords Damage • Electrical resistance • Rock • Cement mortar • Acoustic-emission monitoring

16.1 Introduction

The quantitative evaluation of ongoing damage processes in solid materials is a critical issue due to the treacherous nature of these phenomena, as the slow, gradual deterioration of the material may suddenly degenerate into a catastrophic structural failure. Furthermore, it can be very difficult to macroscopically distinguish a highly damaged volume element from a virgin one, since depth of cracks or inner defects cannot be quantified or identified. It therefore becomes necessary to imagine internal variables representing the deteriorated state of the material, which are directly accessible to measurements.

Around 1500, Leonardo da Vinci was already preoccupied with the characterization of fracture by means of mechanical variables. However, it is only in 1958 that the development of damage mechanics began. In that year, Kachanov

G. Niccolini (✉) • O. Borla • G. Lacidogna • A. Carpinteri
Department of Structural, Geotechnical and Building Engineering,
Politecnico di Torino, Corso Duca degli Abruzzi 24, 10129 Torino, Italy
e-mail: gianni_nicc@hotmail.com

© Springer International Publishing Switzerland 2015   237
A. Carpinteri et al. (eds.), *Acoustic, Electromagnetic, Neutron Emissions from Fracture and Earthquakes*, DOI 10.1007/978-3-319-16955-2_16

published the first paper devoted to a continuous damage variable directed towards modeling the deterioration of materials prior to macroscopic failure [1–4]. In literature research studies investigated on the feasibility of making experimentally accessible the damage variable introduced in damage mechanics [5–7].

Among all damage investigation techniques, the interest is focused on the non-destructive ones, like ultrasonic testing, for which no direct access to the specimen interior is needed. One of the most flexible non-destructive testing methods is the acoustic-emission (AE) technique, as it can be applied for real time damage detection both on laboratory specimens under loading tests and also in situ on in service-structures [8–10]. The AE technique exploits the spontaneous release of elastic energy within stressed materials due to irreversible damage phenomena such as propagation of micro-cracks, corrosion and degradation. The AE signals are detected by suitable sensors, directly coupled on the surface of the structure under test, to point out the presence and location of evolving damage processes.

In recent years an increasing experimental evidence of energy emission of different forms from solid-state fractures has been found. A number of laboratory studies have revealed also the existence of electromagnetic emission (EME) during fracture experiments carried out on a wide range of materials, such as dry and wet granite specimens under triaxial deformation [11, 12]. The EME during rock failure is analogous to the anomalous electromagnetic activity observed before some major earthquakes. Then, the interest in determining the in situ properties of rock and cement mortar based materials, as well as their stress state and how these quantities are changing with time, has been extended to the electrical properties with particular reference to the electrical resistance [13–20].

For relatively dry materials it seems reasonable that the electrical resistance should increase due to microcrack opening, which causes breakdown of the existing conductive network within the material. Since microcracks opening generates acoustic emissions, a correlation between electrical resistance variations and bursts of AE activity is expected [21].

The objective of the present study is to investigate the correlation of electrical resistance with accumulating damage measured by AE in air-dry surface cement mortar and rock specimens subjected to fracture tests. Here, the application of the well-known AE technique aims to verify the reliability of the electrical resistance measurement, which would enable damage monitoring with simple and inexpensive equipment.

16.2   Experimental Results and Data Analysis

A schematic diagram of the equipment used in conducting the experimental study is shown in Fig. 16.1. The testing materials are one cylindrical green in Luserna stone specimen, a metamorphic rock deriving from a granitoid protolith (diameter 52 mm, height 50 mm), and one prismatic cement mortar specimen (section 40×40 mm^2, height 160 mm) enriched with about 10 % in weight of iron oxide.

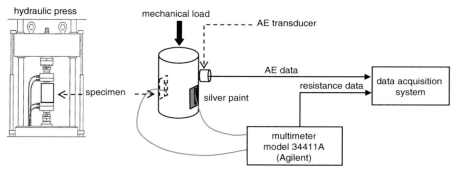

Fig. 16.1 Schematic representation of the experimental set-up

Table 16.1 Chemical composition of mortar enriched with iron oxide and Luserna stone

Mortar		Luserna stone	
Element	Weight (%)	Element	Weight (%)
SiO_2	59.7	SiO_2	72.0
CaO	21.4	Al_2O_3	14.4
Fe_2O_3	8.4	K_2O	4.1
Al_2O_3	3.3	Na_2O	3.7
SO_3	1.1	CaO	1.8
K_2O	1.0	FeO	1.7
MgO	0.7	Fe_2O_3	1.2
Na_2O	0.4	Other oxides	1.1
Other oxides	4.0		

Both materials are characterized by a high percentage of SiO_2, therefore an increase of 10 % in weight of Fe_2O_3 improves the electrical conductivity of the mortar specimen, being the electrical resistivity of Fe_2O_3 five orders of magnitude smaller than SiO_2 (10^9 against 10^{14} Ω cm). The chemical analysis of mortar and Luserna stone are reported in Table 16.1.

Both specimens were subjected to uniaxial compression till failure at constant displacement rate of $1\,\mu m$ s^{-1} for Luserna stone and $2\,\mu m$ s^{-1} for mortar, using a MTS (Mechanical Testing System) servo-hydraulic press with a maximum capacity of 1000 kN equipped with electronic control.

The method which was used in determining the resistance of the specimens is the two-electrode technique. During compression tests, DCR (Direct Current Electrical Resistance) measurements were conducted over the specimens width using an Agilent 34411A multimeter capable of measuring resistances up to 1 GΩ. The resistance was measured through the constant voltage method. In this method, a constant voltage source, V, was placed in series with the multimeter and the unknown resistance R. The current caused to flow through the specimen by the voltage is measured and the resistance is given by the Ohm law [13–21].

The electrical contacts, located symmetrically with respect to each other on opposite faces of the specimen, were made through brass screws in conjunction with copper wires. The primary consideration using direct-current methods is that of obtaining a minimum amount of contact resistance between the electrodes and the specimen. As reported in the literature [21], we used silver electrodes deposited on the specimen faces in the form of a conductive paint having an average diameter of 30 mm. Good results may be also obtained with gold or silver electrodes deposited on the specimen in the form of a conductive paint or evaporated onto the samples in a vacuum chamber, but the Authors did not have these methodologies available.

The value R_0 of the electrical resistance for the undamaged specimens was measured at the beginning of each test. Then, the changing value R was monitored until the specimen failure, expressing the change in terms of R/R_0. The resistance R was measured at a sampling rate of 25 Hz. The reported values were obtained by averaging over 4 s, i.e. over 100 samples.

Acoustic emissions were detected by means of a piezoelectric calibrated accelerometer (charge sensitivity 9.20 pC/m s^{-2}) working in the range of few hertz to 10 kHz, in order to study larger fractures which are able to breakdown the existing conductive network within the material.

As a matter of fact, during damage of a brittle material in compression, micro and macro cracks cause mechanical vibrations of frequencies and wave lengths related to the size of the cracks. While at the beginning of the damage, micro cracks —able to produce AE waves in the frequency range comprised between 50 and 500 kHz– are generated, as the failure is approaching, macro cracks are created which involve perturbations characterized by relevant amplitudes and relatively low oscillation frequencies, generally close to 10 kHz. These observations are well explained in two recent papers of the authors [22, 23].

The AE sensor was mounted on the specimen surface using an epoxy resin for the mortar specimen. For the cylindrical granite specimen the sensor was rigidly stud mounted using a specially designed concave adapter. Each AE event was characterized by the time of occurrence and the magnitude $m = \log_{10} (a_{max} / 1 \, \mu m \, s^{-2})$, being a_{max} the peak acceleration on the specimen surface produced by the AE wave. AE data were acquired at the sampling rate of 44.1 kHz, setting the detection threshold to 40 dB in a way that no spurious signals were detected before the beginning of the tests. In post-process FFT (Fast Fourier Transform) signal analysis the mechanical noise of the press was identified and filtered out. Therefore, the time series and the accumulated number of remaining AE events, the applied load and the relative resistance R/R_0 (in semi-logarithmic scale to emphasize the variations) were plotted versus time.

From the load-time diagram of mortar in Fig. 16.2 (*left*), it is observed that linear elasticity is applicable until the specimen failure. This behaviour is referred to as brittle failure. Despite the absence of significant deviations from linear elasticity, the increase in electrical resistance and AE activity during the approach to failure indicates that damage was accumulating within the specimen. The correlation between the occurrence of AE bursts and resistance change is illustrated in Fig. 16.2 (*left*).

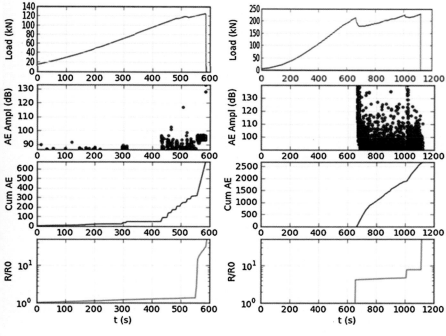

Fig. 16.2 Time plots of: applied load, AE event series, accumulated AE number and electrical resistance variation R/R_0 (in semi-logarithmic scale) for cement mortar (*left*) and Luserna stone specimen (*right*)

Actually, the load-time diagram of the Luserna stone specimen (Fig. 16.2 (*right*)) was decorated by two significant load drops (at 650 and 1000 s), signature of crack advancements and closely correlated with AE bursts (containing events up to 140 dB) and electrical resistance variations. That reveals an ongoing damage process prior to failure.

Figure 16.3, represented in semi-logarithmic scale to emphasize the variations, shows the electrical resistance vs. stress for the two specimens. The stress values (MPa) were simply obtained by dividing the forces (kN) with the nominal cross sectional area of the specimens. As a first hint, it is evident the different electrical resistance behaviour of the two materials during the compression tests, depending on their physical-mechanical properties. Green Luserna granite is a metamorphic rock deriving from a granitoid protolith with a very heterogeneous crystalline structure and porosity. On the other hand the cementitious specimen is an artificial material vibrated and compacted during the manufacturing process, and for these reasons presents more homogeneous characteristics.

Due to the dependence of rocks resistivity on the porosity (also known as Archie's law [24]) different variations in electrical resistivity were observed. The mortar sample shows a constant electrical resistance value up to the failure, whereas Luserna rock is characterized by a small decrease in the electrical resistivity until

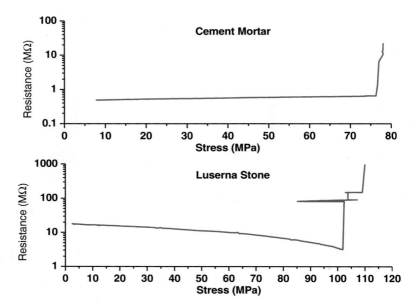

Fig. 16.3 Electrical resistance vs stress for the two specimens

the first stress drop. Moreover, while the mortar specimen is characterized by a significant electrical resistance variation only at the catastrophic failure, for the Luserna stone specimen also sensible electrical resistance changes, caused by internal cracking of the rock, were observed immediately after the consecutive stress-drops.

In particular, after the first stress-drop, a plateau in which the electrical resistance remained almost constant until the 90 % of the peak load is observed. Finally, an intense electrical resistance variation was detected at final failure. These significant variations of the electrical resistance, that precede the final collapse, are clearly precursors of specimen failure, in particular for rock specimens often characterized by stress-drops during the compression tests [22, 23].

16.3 Measurement of Damage by Acoustic Emissions

In continuum mechanics a damage variable D can be defined as the surface density of micro-cracks and cavities in any plane of a representative volume element V. If S is a cross-sectional area (with normal $\mathbf{n}$) of V, and S_0 is the total area of the defect traces on this section, $D_n = S_0/S$ measures the local damage relative to the $\mathbf{n}$-direction; $D_n = 0$ describes the undamaged state, whereas $D_n = 1$ the rupture of V into two parts along the plane $\mathbf{n}$. In case of defects without preferred orientation, damage is completely characterized by a scalar: $D_n = D$, $\forall \mathbf{n}$ [1–4].

Here, we considered damage as given by active defects, which are able to produce acoustic emissions. Thus, considering m the magnitude of the AE signals, the cumulative damage D was expressed in terms of acoustic emissions [5–7]:

$$D \propto \sum_i 10^{m_i} \tag{16.1}$$

Moreover, it has been shown that the magnitude is proportional to the logarithm of the source rupture area s [5–7]:

$$m \propto \log_{10} s \tag{16.2}$$

Combining Eqs. (16.1) and (16.2) shows that D is proportional to the cumulative area of the newly formed microcrack faces, proving the consistency of Eq. (16.1):

$$D \propto \sum_i s_i \tag{16.3}$$

The cumulative damage is normalized to one, so that $D = 1$ represents the maximum damage at the material failure.

16.4   Measurement of Damage by Variations of Electrical Resistance

The electrical resistance R_0 of the undamaged specimen of length l, constant cross-sectional area S and electrical resistivity ρ is expressed as:

$$R_0 = \rho \frac{l}{S} \tag{16.4}$$

The resistance of the specimen experiencing damage during mechanical tests can be expressed as [2, 13, 14]:

$$R = \rho' \frac{l}{S'} = \rho' \frac{l}{S(1 - D)} \tag{16.5}$$

where length changes are neglected, $l = l'$, the damage distribution is assumed to be constant along the length and on the cross-section of the specimen, and the effective current-conducting area S' is identified by the effective load-carrying area $(1 - D) S$.

As the failure stress is approached, the nucleation of microcracks causes more void space in the material and consequently a higher electrical resistivity. This is expressed by the Bridgman law [13, 14, 25, 26], in which the resistivity ρ' is affected by damage only through the relative change of density $(1 - \gamma'/\gamma)$ between the damaged state γ' and the undamaged state γ, and K is a material coefficient:

$$\rho' = \rho\left[1 + K\left(1 - \gamma'/\gamma\right)\right] \tag{16.6}$$

On the other hand, the cumulative damage D is related to the change of density by:

$$D = \left(1 - \gamma'/\gamma\right)^{2/3} \tag{16.7}$$

Equation (16.7) can be easily derived combining the following relations obtained for a spherical cavity of radius r due to an opening microcrack in a spherical volume element of initial radius r_0 and mass M: $\gamma = 3M \ (4\pi \ r_0^3)^{-1}$, $\gamma' = 3M \ [4\pi \ (r_0^3 + r^3)]^{-1}$, $D = \pi r^2 \ [\pi \ (r_0^3 + r^3)]^{-2/3}$ [26].

Combining further Eqs. (16.4), (16.5), (16.6) and (16.7), gives a relation between the resistance ratio R_0/R and the cumulative damage D [13, 14]:

$$\frac{R_0}{R} = \frac{1 - D}{1 + KD^{3/2}} \tag{16.8}$$

Equation (16.8) correctly gives $R = R_0$ when $D = 0$, and $R \to \infty$ when $D = 1$ (infinite resistance at the specimen rupture).

We applied the scaling law expressed in Eq. (16.8) to the quasi-brittle response of Luserna stone, correlating the cumulative damage given in terms of total sum of acoustic emissions by Eq. (16.1) and the measured resistance ratio. The fit yielded $K \approx 55$ which indicates that the electrical resistivity of rocks was strongly dependent on the porosity, as widely reported in literature (Fig. 16.4) [11].

16.5   Measurement of Damage and *b*-value Analysis

The frequency of occurrence of acoustic emissions during a cracking process was related to their magnitude by the Gutenberg-Richter law (originally introduced in earthquake seismology) [27]:

$$\dot{N} = \dot{a}\, 10^{-bm} \tag{16.9}$$

where b, or b-value, and $\dot{a}$ are positive constants, and $\dot{N}$ is the number of acoustic emissions per unit time with magnitude greater than m occurring during the monitoring.

The b-value changes with the different stages of damage process and then it is used to follow damage evolution [5–7]: when the process is dominated by diffused microcracking the b-value is high, while low b-values correspond to macrocrack growth. Therefore, a correlation between the trends of cumulative damage and b-value was expected.

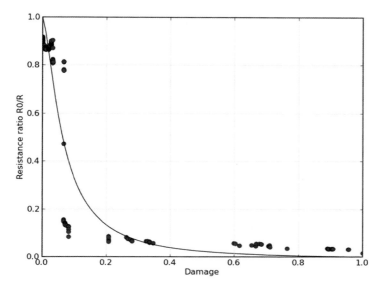

Fig. 16.4 Fitting curve (*continuous line*) correlating the measured resistance ratio R_0/R with cumulative damage D in terms of total sum of acoustic emissions (*blue* points)

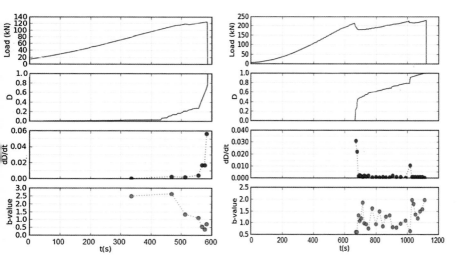

Fig. 16.5 Applied load, cumulative damage D, damage rate $\dot{D}$ and b-value vs. time for mortar (*left*) and Luserna stone (*right*)

The trends of b-value and damage rate $\dot{D}$ for Luserna stone and mortar were calculated using groups of 100 AE events, and then plotted versus time. Figures 16.5 (*left*) and (*right*) show that minima of b-value were correlated with accelerated damage rate, in agreement with other experiments [6].

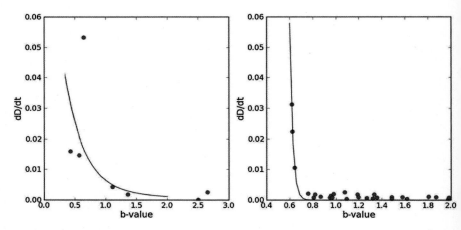

Fig. 16.6 Fitting function (*continuous line*) given by Eq. (16.12) correlating damage rate $\dot{D}$ with b-value, and experimental data (*circles*) for mortar (*left*) and Luserna stone (*right*)

Then, we investigated analytically the relation between the damage variable D and the b-value. Combining Eqs. (16.2), (16.3), (16.4), (16.5), (16.6), (16.7), (16.8), and (16.9) gives [28]:

$$\dot{N} \propto s^{-b} \qquad (16.10)$$

where $\dot{N}$ here represents the number of newly formed cracks per unit time with rupture areas greater than s.

Taking the derivative of Eq. (16.10) gives:

$$d\dot{N} \propto b s^{-b-1} ds \qquad (16.11)$$

Thus, the damage rate $\dot{D}$ was expressed in terms of the rupture area rate $\dot{S}$:

$$\dot{D} \propto \dot{S} = \int_{s_{min}}^{s_{max}} s\, d\dot{N} \propto \int_{s_{min}}^{s_{max}} (-b)\, s^{-b} ds = \frac{b}{b-1}\left(s_{max}^{1-b} - s_{min}^{1-b}\right) \qquad (16.12)$$

where s_{min} and s_{max} represent respectively the smallest and the largest crack area.

Fitting the data with Eq. (16.12) yielded s_{max}/s_{min} equal to 604.2 and 255.1 respectively for Luserna stone and mortar. These values were comparable, or close, to the magnitude ratios ($10^{m_{max}}/10^{m_{min}}$) which were equal to 311.2 and 209.0. This was illustrated in Fig. 16.6 (*left*) and (*right*).

16.6   Conclusions

The experimental information collected during laboratory fracture tests on Luserna stone and cement mortar specimens shows that acoustic emissions and changes in electrical resistance are correlated failure precursors. The result seems to be remarkable for brittle deformation processes, in which abrupt failure occurs with a limited evidence of mechanical degradation.

The distinction between brittle and quasi-brittle deformations is further discussed by verifying for the latter a scaling law between electrical resistance and cumulative damage. With this approach it is shown that the undoubted difficulties in determining the damage level under load by measuring the electrical resistance variation in rocks and cement based materials, due to their low electrical conductivity, are greatly improved when those evaluations are coupled with the damage measurements by AE (see diagram of Fig. 16.4).

Furthermore, the conventional b-value analysis applied to the acoustic-emission data shows that minima of the b-value analysis corresponds to the sudden increase of damage represented by the cumulative damage D. Finally, a second scaling law correlating these two parameters is proposed. This could be tested coupling visual inspections of external macrocracks with AE monitoring.

Acknowledgments   The Authors wish to thank Dr. Alessandro Schiavi of the Istituto Nazionale di Ricerca Metrologica (Torino, Italy) for his technical support.

References

1. Kachanov LM (1986) Introduction to continuum damage mechanics. Martinus Nijhoff, Dordrecht
2. Lemaitre J, Chaboche JL (1990) Mechanics of solid materials. Cambridge University Press, Cambridge
3. Krajcinovic D (1996) Damage mechanics. Elsevier, Amsterdam
4. Turcotte DL, Newman WI, Shcherbakov R (2002) Micro and macroscopic modes of rock fracture. Geophys J Int 152:718–728
5. Cox SJD, Meredith PG (1993) Microcrack formation and material softening in rock measured by monitoring acoustic emission. Int J Rock Mech Min Sci Geomech Abstr 30:11–24
6. Colombo S, Main IG, Forde MC (2003) Assessing damage of reinforced concrete beam using "b-value" analysis of acoustic emission signals. J Mater Civ Eng ASCE 15:280–286
7. Carpinteri A, Lacidogna G, Niccolini G, Puzzi S (2008) Critical defect size distributions in concrete structures detected by the acoustic emission technique. Meccanica 43:349–363
8. Lockner D (1993) The role of acoustic emissions in the study of rock fracture. Int J Rock Mech Min Sci Geomech Abstr 7:883–899
9. Guarino A, Garcimartin A, Ciliberto S (1998) An experimental test of the critical behavior of fracture precursors. Eur Phys J B 6:13–24
10. Niccolini G, Carpinteri A, Lacidogna G, Manuello A (2011) Acoustic emission monitoring of the Syracuse Athena temple: scale invariance in the timing of ruptures. Phys Rev Lett 106:108503

11. Yoshida S, Ogawa T (2004) Electromagnetic emissions from dry and wet granite associated with acoustic emissions. J Geophys Res 109, B09204
12. Triantis D, Vallianatos F, Stavrakas I, Hloupis G (2012) Relaxation phenomena of electric signal emissions from rocks following to abrupt mechanical stress application. Ann Geophys 55:207–212
13. Sun B, Guo Y (2004) High-cycle fatigue damage measurement based on electrical resistance change considering variable electrical resistivity and uneven damage. Int J Fatigue 26:457–462
14. Chen B, Liu J (2008) Damage in carbon fiber-reinforced concrete, monitored by both electrical resistance measurement and acoustic emission analysis. Construct Build Mater 22:2196–2201
15. Stavrakas I, Anastasiadis C, Triantis D, Vallianatos F (2003) Piezo stimulated currents in marble samples: precursory and concurrent with failure signals. Nat Hazards Earth Syst Sci 3:243–247
16. Triantis D, Anastasiadis C, Stavrakas I (2008) The correlation of electrical charge with strain on stressed rock samples. Nat Hazards Earth Syst Sci 8:1243–1248
17. Kyriazopoulos A, Anastasiadis C, Triantis D, Brown JC (2011) Non-destructive evaluation of cement-based materials from pressure-stimulated electrical emission -preliminary results. Construct Build Mater 25:1980–1990
18. Chen G, Lin Y (2004) Stress-strain-electrical resistance effects and associated state equations for uniaxial rock compression. Int J Rock Mech Min Sci 41:223–236
19. Chung DDL (2003) Damage in cement-based materials, studied by electrical resistance measurement. Mater Sci Eng R Rep 42:1–40
20. Wen SH, Chung DDL (2000) Damage monitoring of cement paste by electrical resistance measurement. Cem Concr Res 30:1979–1982
21. Russell JE, Hoskins ER (1969) Correlation of electrical resistivity of dry rock with cumulative damage. The 11th U.S. symposium on rock mechanics, American Rock Mechanics Association, Berkeley
22. Carpinteri A, Lacidogna G, Manuello A, Niccolini G, Schiavi A, Agosto A (2012) Mechanical and electromagnetic emissions related to stress-induced cracks. Exp Tech 36:53–64
23. Lacidogna G, Carpinteri A, Manuello A, Durin G, Schiavi A, Niccolini G, Agosto A (2011) Acoustic and electromagnetic emissions as precursor phenomena in failure processes. Strain 47(2):144–152
24. Archie GE (1942) The electrical resistivity log as an aid in determining some reservoir characteristics. Pet Trans AIME 146:54–62
25. Bridgnman PW (1932) The effect of homogeneous mechanical stress on the electrical resistance of crystals. Phys Rev 42:858
26. Lemaitre J, Dufailly J (1987) Damage measurements. Eng Fract Mech 28:643–661
27. Richter CF (1958) Elementary seismology. Freeman, San Francisco
28. Kanamori H, Anderson DL (1975) Theoretical basis of some empirical relations in seismology. Bull Seismol Soc Am 65:1073–1095

Chapter 17
The Sacred Mountain of Varallo Renaissance Complex in Italy: Damage Analysis of Decorated Surfaces and Structural Supports

Federico Accornero, Stefano Invernizzi, Giuseppe Lacidogna, and Alberto Carpinteri

Abstract Acoustic Emission (AE) is a Non-Destructive Inspection Technique, widely used for monitoring of structural condition of different materials like concrete, masonry and rocks. It utilizes the transient elastic waves produced by each fracture occurrence, which are captured by sensors on the external surface.

The preservation of the mural painting heritage is a complex problem that requires the use of innovative non-destructive investigation methodologies to assess the integrity of decorated artworks without altering their state of conservation. A complete diagnosis of crack pattern regarding not only the external decorated surface but also the internal support is of great importance due to the criticality of internal defects and damage phenomena, that may suddenly degenerate into irreversible failures. The majority of NDT work by introducing some type of energy into the system to be analysed. On the other hand, in AE tests the input is the mechanical stress inside the material itself during the damage evolution, so that no perturbation is induced and the integrity of the system can be guaranteed. By monitoring the support of a decorated surface by means of the AE technique, it becomes possible to detect the occurrence and evolution of surface vs. support separation and stress-induced cracks.

The aim of this study is to reveal by means of the AE technique the damage evolution in the support of the decorated surfaces of the Renaissance Complex "Sacri Monti di Varallo" (Piedmont, Italy) and to utilize the collected data coming from the "in situ" monitoring in order to preserve the artworks from seismic risk and possible collapses due to earthquake actions.

Keywords Acoustic emission • Non-destructive testing • Historical heritage preservation • Damage assessment • Seismic risk

F. Accornero (✉) • S. Invernizzi • G. Lacidogna • A. Carpinteri
Department of Structural, Geotechnical and Building Engineering, Politecnico di Torino, Corso Duca degli Abruzzi 24, 10129 Torino, Italy
e-mail: federico.accornero@polito.it

© Springer International Publishing Switzerland 2015
A. Carpinteri et al. (eds.), *Acoustic, Electromagnetic, Neutron Emissions from Fracture and Earthquakes*, DOI 10.1007/978-3-319-16955-2_17

249

17.1 Introduction

The Sacred Mountain of Varallo is the most ancient Sacred Mountain of Piedmont and Lombardy and it is located among the green of the forests at the top of a rocky spur right above the city of Varallo (Fig. 17.1). It consists in 45 Chapels, some of which are isolated, while others are part of monumental groups. They contain over 800 life-size wooden and multicoloured terracotta statues, which represent the Life, the Passion and the Death of Christ.

The Sacred Mountain of Varallo is the work of two great churchmen and of a number of artists headed by Gaudenzio Ferrari [1]. The two churchmen are: the Franciscan Friar, Blessed Bernardino Caimi and St. Charles Borromeo Archbishop of Milan. At Varallo, Fra Bernardino Caimi put into practice the idea that he had been turning over in his mind during his stay in the Holy Land. His aim was to erect buildings that would recall the "Holy Places" of Palestine. Those places evoke the characteristic monuments of Christ's stay on earth (the Stable at Bethlehem, the House in Nazareth, the Last Supper, the Calvary and Holy Sepulchre). He began his work in 1491 and carried on with it as long as he lived (until the end of 1499), assisted by Gaudenzio Ferrari who continued his idea and decorated a number of chapels with frescos and statues in wood and terracotta.

St. Charles Borromeo appreciated the work already done when he paid a visit to the Sacro Monte in 1578 and, giving the place the appropriate name of "New Jerusalem", made it more widely known among his contemporaries. Returning there at the end of October 1584, he decided to develop the original idea by building new chapels, which would illustrate the life of Jesus more completely.

For the great Bishop of Milan it was an effective mean of his time, giving the population greater religious fervour and protecting them from the heresies that threatened Northern Italy. He utilized the project for the rearrangement of the Sacred Mountain drawn up by the architect Galeazzo Alessi in 1592 and, adapting it to his own plans, gave instructions for the resumption of work. The work continued until 1765.

Today the Sacred Mountain of Varallo continues to be a school of Christian truth and life, while at the same time it is the most precious treasury of art in the Valsesia Valley.

Fig. 17.1 The sacred mountain of Varallo: view of the square of tribunals

Fig. 17.2 External view of Chapel XVII (**a**); internal view of the mount tabor installation (**b**)

As the first object of investigation with the AE technique, the Chapel XVII of the Sacred Mountain of Varallo is chosen (Fig. 17.2). This Chapel houses the scene of the Transfiguration of Christ, who appeared to the Apostles at the foot of the mountain, in radiant light between Elijah and Moses.

This Chapel was also foreseen in the "Book of Mysteries" by Galeazzo Alessi (1565–1569) [1, 2], from which the group of sculptures located high on the mountain take inspiration. The relative foundations were already begun in 1572, but the chapel was not completed until the 1660s.

17.2   Damage Analysis of the Decorated Surface Structural Supports by the AE Technique

The preservation of the mural painting heritage is a complex problem that requires the use of non-destructive investigation methodologies to assess the integrity of decorated artworks without altering their state of conservation. The physical-chemical decay and the damage evolution of materials constituting the decorated surfaces and the structural supports can be caused by infiltrations of water, thermo-elastic stresses, or seismic and environmental vibrations. The physical-chemical degradation has to be dealt with Materials Science and Chemical Engineering techniques [3]. On the other hand, the instability and the dynamic behaviour of the decorated surfaces, induced also by seismic and environmental vibrations, can be investigated by the Acoustic Emission technique (AE), using monitoring systems to control continuously and simultaneously different structural supports [4].

The data collected during the in situ experimental tests can be interpreted with Fracture Mechanics models and methodologies [5–9].

A complete diagnosis of crack pattern regarding not only the external decorated surface but also the internal support is of great importance due to the criticality of internal defects and damage phenomena, which may suddenly degenerate into irreversible failures [10, 11].

The majority of non-destructive techniques work by introducing some type of energy into the system to be analyzed. On the other hand, in AE tests the input is the mechanical energy release generated by the material itself during the damage evolution, so that no perturbation is induced and the integrity of the system may be guaranteed. By monitoring the support of a decorated surface by means of the AE technique, it becomes possible to detect the occurrence and evolution of surface vs. support detachments and of stress-induced cracks.

Cracking, in fact, is accompanied by the emission of elastic waves, which propagate through the bulk of the material. These waves can be received and recorded by piezoelectric transducers (PZT) applied to the external surface of the artwork support.

Objective of the present research is to use the AE technique to assess the support of the decorated mural surfaces, developing the application aspects of this technique, which has been widely studied from a theoretical and experimental point of view by the authors for the safeguard of civil and historical buildings [12–14].

In a first stage, it will be essential to recognize the artwork to be monitored, its conservation state and the severity of its conditions at the beginning of the monitoring and restoration processes. The AE technique makes it also possible to detect and localize the presence of cracks and analyze the damage evolution in supports such as decorated masonry walls and vaults [15].

From a physical point of view, investigations with AE technique of microfracture processes have revealed power-law distributions and critical phenomena [16–18] characterized by intermittency of AE event avalanches [19], fractal distributions of AE event locations, and complex space-time coupling [20]. Localization of cracks distribution within the specimen volume by means of AE technique has permitted to confirm that the energy is dissipated over preferential bands and surfaces during the damage evolution [21]. Moreover, after the work of Scholz on the b-value [22], Aki [23] was the first to show in an empirical way that the seismic b-value is related to the fractal dimension D, and that usually $2b = D$. This assumption, and its implication with the damage release rate and time dependent mechanisms, both at the laboratory and at the Earth's crust scale, has been pointed out in [24, 25].

In this framework, Acoustic Emission data have been interpreted by means of statistical and fractal analysis, considering the multiscale aspect of cracking phenomena [5]. Consequently, a multiscale criterion to predict the damage evolution has been formulated.

Recent developments in fragmentation theories [26] have shown that the energy W during microcrack propagation is dissipated over a fractal domain comprised between a surface and the specimen volume V.

The following *size-scaling* law has been assumed during the damage process:

$$W \propto N \propto V^{D/3} \tag{17.1}$$

In Eq. (17.1) D is the so-called fractal exponent comprised between 2 and 3, and N is the cumulative number of AE signals that the structure provides during the damage monitoring.

The authors have also shown that energy dissipation, as measured with the AE technique during the damaging process, follows the *time-scaling* law [27]:

$$W \propto N \propto t^{\beta_t} \tag{17.2}$$

where β_t is the time-scaling exponent for the dissipated energy in the range $(0, 3)$ and N is the number of AE signals.

By working out the exponent β_t from the data obtained during the observation period, we can make a prediction on the structure's stability conditions: if $\beta_t < 1$ the structure evolves toward stability conditions; if $\beta_t \approx 1$ the process is metastable; if $\beta_t > 1$ the process is unstable.

Moreover, a statistical interpretation to the variation in the b-value during the evolution of damage detected by AE has been proposed, which is based on a treatment originally proposed by Carpinteri and co-workers [28, 29]. The proposed model captures the transition from the condition of diffused criticality to that of imminent failure localisation.

By analogy with seismic phenomena, in the AE technique the magnitude may be defined as follows:

$$m = Log_{10}A_{max} + f(r) \tag{17.3}$$

where A_{max} is the amplitude of the signal expressed in volts, and $f(r)$ is a correction factor taking into account that the amplitude is a decreasing function of the distance r between the source and the sensor.

In seismology the empirical Gutenberg-Richter's law [30]:

$$Log_{10}N(\geq m) = a - bm \text{ or } N(\geq m) = 10^{a-bm} \tag{17.4}$$

expresses the relationship between magnitude and total number of earthquakes with the same or higher magnitude in any given region and time period, and it is the most widely used statistical relation to describe the scaling properties of seismicity. In Eq. (17.4), N is the cumulative number of earthquakes with magnitude $\geq m$ in a given area and within a specific time range, whilst a and b are positive constants varying from a region to another and from a time interval to another. Equation (17.4) has been used successfully in the AE field to study the scaling laws of AE wave amplitude distribution. This approach evidences the similarity between structural damage phenomena and seismic activities in a given region of the Earth's crust, extending the applicability of the Gutenberg-Richter's law to Structural Engineering. According to Eq. (17.4), the b-value changes systematically at different times in the course of the damage process and therefore can be used to estimate damage evolution modalities.

Equation (17.4) can be rewritten in order to draw a connection between the magnitude m and the size L of the defect associated with an AE event. By analogy with seismic phenomena, the AE crack size-scaling entails the validity of the relationship:

$$N(\geq L) = cL^{-2b} \tag{17.5}$$

where N is the cumulative number of AE signals generated by source defects with a characteristic linear dimension $\geq L$, c is a constant of proportionality, and $2b = D$ is the fractal dimension of the damage domain.

It has been evidenced that this interpretation rests on the assumption of a dislocation model for the seismic source and requires that $2.0 \leq D \leq 3.0$, i.e., the cracks are distributed in a fractal domain comprised between a surface and the volume of the analysed region [23, 31, 32].

The cumulative distribution (17.4) is substantially identical to the cumulative distribution proposed by Carpinteri [28], which gives the probability of a defect with size $\geq L$ being present in a body:

$$P(\geq L) \propto L^{-\gamma} \tag{17.6}$$

Therefore, the number of defects with size $\geq L$ is:

$$N^*(\geq L) = cL^{-\gamma} \tag{17.7}$$

where γ is a statistical exponent measuring the degree of disorder, i.e. the scatter in the defect size distribution, and c is a constant of proportionality. By equating distributions (17.4) and (17.6) it is found that: $2b = \gamma$. At the collapse, the size of the maximum defect is proportional to the characteristic size of the structure. As shown by Carpinteri and co-workers [29], the related cumulative defect size distribution (referred to as self-similarity distribution) is characterized by the exponent $\gamma = 2.0$, which corresponds to $b = 1.0$. It was also demonstrated by Carpinteri [29] that $\gamma = 2.0$ is a lower bound which corresponds to the minimum value $b = 1.0$, observed experimentally when the load bearing capacity of a structural member has been exhausted.

Therefore, by determining the b-value it is possible to identify the energy release modalities in a structural element during the monitoring process. The extreme cases envisaged by Eq. (17.1) are $D = 3.0$, which corresponds to the critical condition $b = 1.5$, when the energy release takes place through small defects homogeneously distributed throughout the volume, and $D = 2.0$, which corresponds to $b = 1.0$, when energy release takes place on a fracture surface. In the former case diffused damage is observed, whereas in the latter two-dimensional cracks are formed leading to the separation of the structural element.

In the following, the data obtained from the monitoring are mainly interpreted by Eqs. (17.2) and (17.4).

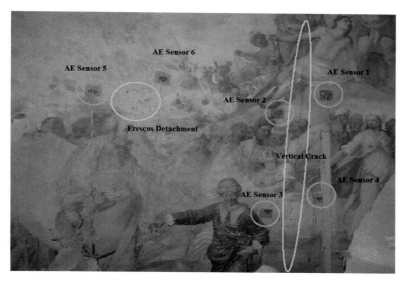

Fig. 17.3 Chapel XVII: view of the monitored damages and position of the AE sensors

As regards the structural integrity, Chapel XVII shows a vertical crack of about 3.00 m in length and a detachment of frescos both on the North wall, which are the object of the present monitoring campaign by means of AE. Six AE sensors are employed to monitor the damage evolution of the structural support of the decorated surfaces of the Chapel XVII: four are positioned around the vertical crack while two are positioned near the frescos detachment (Fig. 17.3). For the sensor pasting on decorated surfaces, a suitable methodology is applied.

The equipment used for signal acquisition and processing consists of six USAM units. Each of these units analyses in real time, and transmits to a PC, all the characteristic parameters of an ultrasonic event. In this manner, each AE event is identified by a progressive number and characterised by a series of data giving the amplitude and time duration of the signals and the number of oscillations. Moreover, the absolute acquisition time and signal frequency are also given, so that, through an analysis of signal frequency and time of arrival at the transducers, it is possible to identify the group of signals belonging to the same AE event and to localise it. Each unit is equipped with a pre-amplified wide-band piezoelectric sensor (PZT), which is sensitive in a frequency range between 50 kHz and 800 kHz. The signals acquisition threshold can be set in a range between 100 μV and 6.4 mV.

Moreover, Chapel XVII shows another vertical crack on the South wall, symmetric to the previous one with respect to the pronao of the building. The monitoring period of the structural supports of the chapel began on April 28, 2011 and ended on June 4, 2011, it lasted about 900 h. The results obtained by the application of the AE sensors are presented in Figs. 17.4 and 17.5. For the vertical crack (Fig. 17.4), approximately 550 AE signals were analyzed, while for the frescos

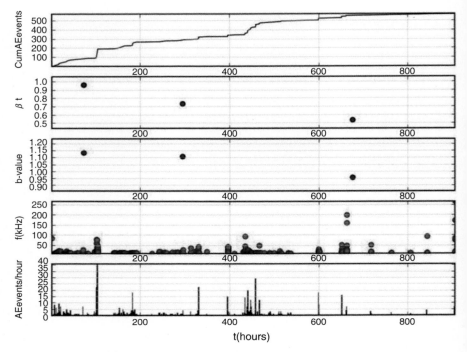

Fig. 17.4 Chapel XVII: AE from vertical crack monitoring

detachment (Fig. 17.5) 1,200 AE signals were considered. The cumulated AE signals, AE rates, β_t parameters and b-values are shown in Figs. 17.4 and 17.5. In particular, to calculate the β_t parameters and the b-values represented in Fig. 17.4, about 200 data for time were used, while about 400 data for time were used for the same parameters reported in Fig. 17.5. More specifically, the three b-values calculated for the vertical crack (Fig. 17.4) are shown in Fig. 17.6 with the corresponding coefficients of determination R^2.

As can be seen from Fig. 17.4, the vertical crack monitored on the North wall of the chapel presents a stable condition during the acquisition period ($0.5 < \beta_t < 1.0$) and a distribution of cracks on a surface domain is clearly proved by the b-value in the range (0.95, 1.15). In this case, it is interesting to note that, since the monitored wall is of large dimensions, a b-value approaching to 1 does not imply a substantial loss of load-bearing capacity of the entire wall, but rather the coalescence of the microcracks along the surface of the vertical crack, which can continue to advance without substantially compromising the structure bearing capacity. Further evidence for the presence of the crack is offered by the low frequency signals registered (< 200 kHz): as a matter of fact, considering the velocity as a constant and applying the Lamb ratio [33], the wavelength needs to be larger than the size of the maximum inhomogeneity in order for the wave to pass through without significant modifications in its waveform. It is reasonable to assume that for a high frequency wave it is possible only to propagate

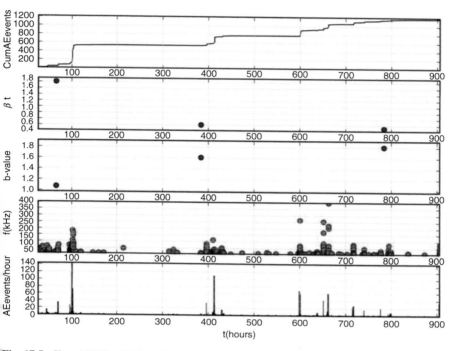

Fig. 17.5 Chapel XVII: AE from frescos detachment monitoring

through a small inhomogeneity; on the contrary for a low frequency wave it is possible also to propagate through a large inhomogeneity [5, 34]. Concerning the monitored frescos detachment (Fig. 17.5), the decorated surface tends to evolve towards metastable conditions ($0.5 < \beta_t < 1.8$) and the signals acquired show high frequency characteristics (<400 kHz): therefore a distribution of microcracks in the volume is assumed for the analysed region.

17.3   Laboratory AE Measurements Due to Capillary Rise in Mortar

A preliminary experimental test has been set up in the laboratory to find out evidences for the recorded AE. The purpose is to find a correlation between specific analyzed AE activity and damage of the decorated surface due to different water cycle stages (e.g. immersion, drying and cooling) linked to the phenomenon of capillary rise. A mortar specimen was monitored by means of AE during 60 min of immersion in mineral water (Fig. 17.7a) with USAM unit above mentioned. The problem is at the present time practically unexplored, although some Authors reported analogous AE activity recorded in similar conditions [35].

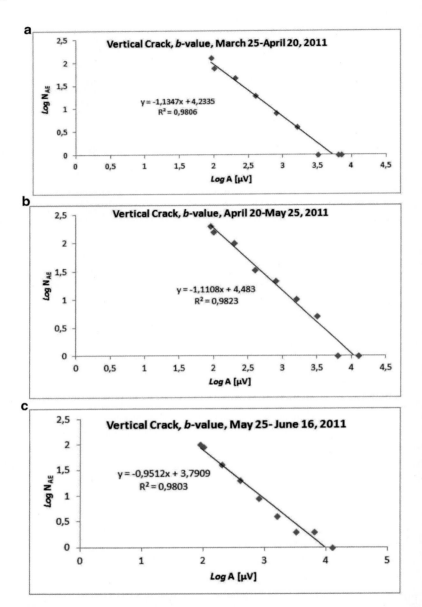

Fig. 17.6 Chapel XVII: *b*-values computed for the vertical crack with the corresponding coefficients of determination R^2; from March 25 to April 20, 2011 (**a**); from April 20 to May 25, 2011
(**b**); from May 25 to June 16, 2011 (**c**)

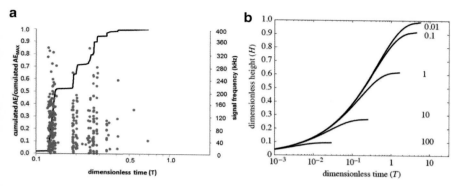

Fig. 17.7 Cumulated number of AE events during mortar immersion in mineral water with AE Frequencies (**a**); dimensionless diagram of the capillary rising. Each curve refers to the dimensionless quantity α, which is a function of the sample thickness, sorptivity and potential evaporation of the micro-environment [36] (**b**)

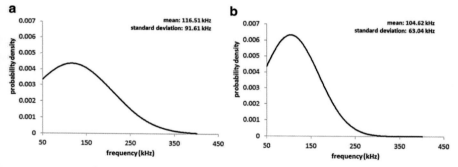

Fig. 17.8 Normal distribution of frequencies registered during mortar immersion test (**a**); normal distribution of frequencies registered during frescos detachment monitoring (**b**)

The first evidence is that the evolution of the cumulated AE resembles the kinetics of the capillary rising in the transient regime (Fig. 17.7b): this underlies how the two phenomena are correlated. We could see also that the frequency field of AE events registered during the mortar immersion test (Fig. 17.7a) reminds the frequency field detected from frescos detachment (Fig. 17.5); a probable reason of the decorated surface deterioration may be found in moisture diffusion through the structural support. The normal distribution of the frequencies obtained during the laboratory test (Fig. 17.8a) shows a dispersion wider than the AE frequency distribution detected "in situ" (Fig. 17.8b). As a matter of fact, the AE signals due to frescos detachment are related to a limited period of damage within the whole life-time of the structure, while the laboratory test investigates the mortar specimen behavior starting from a totally dry condition. Nevertheless, both results show signals in a higher frequency range compared to that produced by the monitored vertical crack (Fig. 17.4).

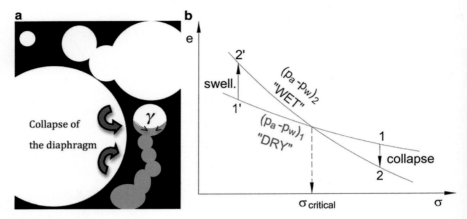

Fig. 17.9 Scheme of the collapse of diaphragm between pores of different diameter due to capillary surface tension γ (**a**); dilatation *e* of an hydrated swelling clay-like material below the critical stress σ_{cr} (**b**): "p_a" means air pressure in the pores, while "p_w" water pressure in the pores

More generally, such a deterioration process characterized by AE activity, could be due to a sort of local rupture of the rigid skeleton of the mortar structure, necessarily leading to microcrack development or to an increase in the number of voids.

Two possible main mechanisms can be assumed, namely due to chemical or mechanical processes. From the chemical point of view, the local damage could be caused by the presence of swelling salts in the mortar components. The crystallization of salts inside the mortar pores is commonly known as *crypto-florescence*. A typical case is sodium sulphate crystallization. A reversible solubility reaction determines the amount of *mirabilite* ($Na_2SO_4 \cdot 10H_2O$) in equilibrium with *thenardite* (Na_2SO_4), as a function of relative humidity and temperature. As the content of water increases, the percentage of *mirabilite* growths, leading to swelling and damage, due to the lower density of *mirabilite* with respect to *thenardite* [35].

From the mechanical point of view, the rise of water into the porous mortar is influenced by the complex distribution of voids of different size and different degree of connection. Therefore, the presence of water in connected network of small pores (Fig. 17.9a), nearby larger empty pores, can realize large unbalanced forces (due to the surface tension γ), which are able to trigger the collapse of the diaphragm. Local ruptures of such kind can be found in the literature about MEMS device [37].

Finally, the presence of very small porosity in the mortar (in the range between 10 and 10^5 nm) can be at the origin of a swelling behavior similar to the one displayed by unsaturated clay-like materials [38]. In this case, the increase in humidity, if the state of stress is below the critical stress, is accompanied by an increase in the volume (Fig. 17.9b). This phenomenon could occur locally in mortar, and the restrained expansion may lead to damage and AE activity.

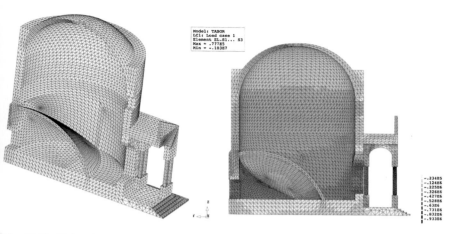

Fig. 17.10 Finite element mesh (**a**); principal compression stress contour (**b**)

17.4 Finite Element Modeling

The chapel was discretized exploiting symmetry with three-dimensional linear pyramid elements, accounting for the accurate geometry of the stone masonry structure. The shape of the cylindrical chapel and of the above emi-spherical dome are precisely discretized, taking into account the various apertures, the internal vault supporting the Mount Tabor installation, and the outside pronao with columns. On the contrary, the wooden roof structure was considered only as an external load. The mesh of the structure is shown in Fig. 17.10a. The finite element model is discretized using 15,400 nodes, connected by 64,200 elements, and is characterized by 43,064 degrees of freedom. The elastic properties assumed for masonry and the density were respectively equal to: $E = 2E + 9$ Pa; $\nu = 0.3$; $\rho = 20$ kN/m^3. The elastic analysis, performed with the commercial finite element code DIANA® [39] allows for a preliminary assessment of the structure. Figure 17.10b shows the contour diagram of the principal compressive stress. In general, the level of compressive stress is rather low, compared with the expected strength of the stone masonry, and almost everywhere lower than 1 MPa. Nevertheless, the compression stress in the external columns of the pronao is greater than or equal to 1.03 MPa. The external columns are also subjected to the environmental degradation of the stone.

Figure 17.11a shows the contour diagram of the principal tensile stress, reported on the deformed configuration of the structure. The tensile stresses calculated on the internal wall of the chapel justify the presence of the two symmetric dominant cracks. Figure 17.11b shows the deformed configuration of the structure compared to the initial shape. The deformation clearly shows the opening mechanism due to the effect of the pronao, as well as to the thrust of the internal vault supporting the mount Tabor installation. A more detailed mechanical characterization of the masonry is currently under development to perform the subsequent nonlinear analysis [40].

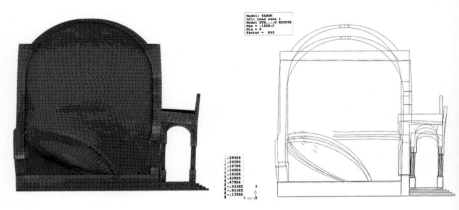

Fig. 17.11 Principal tensile stress contour (**a**); deformed configuration (**b**)

17.5 Conclusions

Chapel number 17, dedicated to the Mount Tabor episode of Christ Transfiguration, of the Sacred Mountain of Varallo shows some structural concern due to cracking and degradation of the high valuable frescos, which tend to detach from the masonry support.

In order to assess the evolution of the phenomena, the results of the Acoustic Emission monitoring program of the Chapel have been provided, together with a structural finite element simulation.

The Finite Element analysis provided the basic mechanism that is taking place in the chapel, which caused two main cracks in the internal walls.

The results of the monitoring show that the large cracks are stable, while the process of detachment of the frescos is evolving cyclically. It seems that the frescos degradation could be mainly related to the diffusion of moisture in the mortar substrate. Some preliminary laboratory tests also confirm that Acoustic Emissions are recorded in mortar samples subjected to moisture diffusion.

References

1. De Filippis E (2009) Sacro Monte of Varallo. Tipolitografia di Borgosesia, Borgosesia
2. Alessi G (1974) Libro dei Misteri. Anastatic Copy. Bologna
3. De Filippis E, Tulliani JM, Sandrone R, Scarzella P, Palmero P, Lombardi SC, Zerbinatti M (2005) Analisi degli intonaci della Cappella del Calvario al Sacro Monte di Varallo. ARKOS 12:38–45
4. Grosse CU, Ohtsu M (2008) Acoustic emission testing. Springer, Heidelberg
5. Carpinteri A, Lacidogna G, Pugno N (2007) Structural damage and life-time assessment by acoustic emission monitoring. Eng Fract Mech 74:273–289
6. Carpinteri A, Lacidogna G, Puzzi S (2009) From criticality to final collapse: evolution of the b-value from 1.5 to 1.0. Chaos Solitons Fractals 41:843–853

7. Carpinteri A, Lacidogna G, Accornero F, Mpalaskas AC, Matikas TE, Aggelis DG (2013) Influence of damage in the acoustic emission parameters. Cem Concr Compos 44:9–16
8. Botvina LR (2011) Damage evolution on different scale levels. Phys Solid Earth 47:859–872
9. Botvina LR, Shebalin PN, Oparina IB (2001) A mechanism of temporal variation of seismicity and acoustic emission prior to macrofailure. Dokl Phys 46:119–123
10. Anzani A, Binda L, Carpinteri A, Lacidogna G, Manuello A (2008) Evaluation of the repair on multiple leaf stone masonry by acoustic emission. Mater Struct 41:1169–1189
11. Carpinteri A, Invernizzi S, Lacidogna G (2009) Historical brick-masonry subjected to double flat-jack test: acoustic emissions and scale effects on cracking density. Construct Build Mater 23:2813–2820
12. Anzani A, Binda L, Carpinteri A, Invernizzi S, Lacidogna G (2010) A multilevel approach for the damage assessment of historic masonry towers. J Cult Herit 11:459–470
13. Carpinteri A, Invernizzi S, Lacidogna G, Manuello A, Binda L (2008) Stability of the vertical bearing structures of the Syracuse Cathedral: experimental and numerical evaluation. Mater Struct 42(7):877–888
14. Invernizzi S, Lacidogna G, Manuello A, Carpinteri A (2011) AE monitoring and numerical simulation of a 2-span model masonry arch bridge subjected to pier scour. Strain 47 (2):158–169
15. Carpinteri A, Invernizzi S, Lacidogna G (2007) Structural assessment of a XVII century masonry vault with AE and numerical techniques. Int J Archit Herit 1(2):214–226
16. Garcimartin A, Guarino A, Bellon L, Ciliberto S (1997) Statistical properties of fracture precursors. Phys Rev Lett 79:3202
17. Sethna JP, Dahmen KA, Myers CR (2001) Crackling noise. Nature 410:242
18. Lu C, Mai Y-W, Shen Y-G (2005) Optimum information in crackling noise. Phys Rev Lett E 72:027101
19. Weiss J, Marsan D (2003) Three-dimensional mapping of dislocation avalanches: clustering and space/time coupling. Science 299:89–92
20. Turcotte DL, Shcherbakov R (2003) Damage and self-similarity in fracture. Theor Appl Fract Mech 39
21. Carpinteri A, Lacidogna G, Puzzi S (2008) Prediction of cracking evolution in full scale structures by the b-value analysis and Yule statistics. Phys Mesomech 11:260–271
22. Scholz CH (1968) The frequency-magnitude relation of microfracturing in rock and its relation to earthquakes. Bull Seismol Soc Am 58:399–415
23. Aki A (1981) A probabilistic synthesis of precursor phenomena. In: Simpson DW, Richards PG (eds) Earthquake prediction: an international review, 4th edn. AGU, Washington, DC, pp 566–574
24. Main IG (1992) Damage mechanics with long-range interactions: correlation between the seismic b-value and the two-point correlation dimension. Geophys J Int 111:531–541
25. Carpinteri A, Lacidogna G, Pugno N (2006) Richter's laws at the laboratory scale interpreted by acoustic emission. Mag Concr Res 58:619–625
26. Carpinteri A, Pugno N (2002) Fractal fragmentation theory for shape effects of quasi-brittle materials in compression. Mag Concr Res 54:473–480
27. Carpinteri A, Lacidogna G, Pugno N (2005) Time-scale effects on acoustic emission due to elastic waves propagation in monitored cracking structures. Phys Mesomech 8:77–80
28. Carpinteri A (1994) Scaling laws and renormalization groups for strength and toughness of disordered materials. Int J Solids Struct 31:291–302
29. Carpinteri A, Lacidogna G, Niccolini G, Puzzi S (2008) Critical defect size distributions in concrete structures detected by the acoustic emission technique. Meccanica 43:349–363
30. Richter CF (1958) Elementary seismology. W.H. Freeman, San Francisco/London
31. Rundle JB, Turcotte DL, Shcherbakov R, Klein W, Sammis C (2003) Statistical physics approach to understanding the multiscale dynamics of earthquake fault systems. Rev Geophys 41:1–30

32. Turcotte DL (1997) Fractals and chaos in geology and geophysics. Cambridge University Press, New York
33. Lamb H (1917) On waves in an elastic plate. Proc R Soc A 93:114–128
34. Landis EN, Shah SP (1995) Frequency-dependent stress wave attenuation in cement-based materials. J Eng Mech 121:737–743
35. Grossi CM, Esbert RM, Suarez del Rio LM, Montoto M, Laurenzi-Tabasso M (1997) Acoustic emission monitoring to study sodium sulphate crystallization in monumental porous carbonate stones. Stud Conserv 42:115–125
36. Hall C, Hoff WD (2007) Rising damp: capillary rise dynamics in walls. Proc R Soc A 463:1871–1884
37. Mastrangelo CH, Hsu CH (1993) Mechanical stability and adhesion of microstructures under capillary forces – part II: experiments. J Micro-electromechanical Syst 2(1):44–55
38. Sheng D, Gens A, Fredlund DG, Sloan SW (2008) Unsaturated soils: from constitutive modelling to numerical algorithms. Comput Geotech 35:810–824
39. Maine J (2012) DIANA. Finite element analysis user's manual release 9.4.4. TNO DIANA bv, Delft
40. Carpinteri A, Lacidogna G, Invernizzi S, Accornero F (2013) The sacred mountain of Varallo in Italy: seismic risk assessment by acoustic emission and structural numerical models. Sci World J 2013:170291

Printed by Printforce, the Netherlands